Deep
Blue

Also by Steve Backshall

Expedition
Looking for Adventure
Mountain: A Life on the Rocks

Deep Blue

My Ocean Journeys

STEVE BACKSHALL

WITNESS
BOOKS

1

Witness Books, an imprint of Ebury Publishing
20 Vauxhall Bridge Road
London SW1V 2SA

Witness Books is part of the Penguin Random House group of companies
whose addresses can be found at global.penguinrandomhouse.com

Penguin
Random House
UK

Verse from the poem 'Sea-Fever', from the collection Sea-Fever: Selected
Poems by John Masefield, reprinted by kind permission of Carcanet Press

Image credits: Steve Backshall, Steve Backshall, Duncan Brake,
Simon Enderby, Keith Partridge, Scubazoo, Deron Verbeck and
Matthew Wright. Illustrations by Doug Mackay-Hope

First published by BBC Books in 2023
Paperback edition published in 2024

www.penguin.co.uk

A CIP catalogue record for this book is available from the British Library

ISBN 9781529144116

Printed and bound in Great Britain by Clays Ltd, Elcograf S.p.A.

The authorised representative in the EEA is Penguin Random House Ireland,
Morrison Chambers, 32 Nassau Street, Dublin D02 YH68

To Logan, Kit and Bo,
for the life of exploration you have before you.

Contents

Prologue

For most of history, man has had to fight nature to
survive; in this century he is beginning to realise
that, in order to survive, he must protect it.

Jacques-Yves Cousteau

The den was a sliver of a hole beneath a fridge-sized boulder,
the welcome mat a litter of king crab carcasses, carapaces like
spiky dinner plates, broken claws once stout enough to clip off a
human thumb.

It was our fifth day of diving in the chilly waters off Vancouver
Island, waiting outside this same den for the inhabitant to show
itself. Less scrupulous voyeurs have pumped washing-up liquid or
engine fuel into the cracks to force an occupant to emerge. We
were sticking to ethical means: a raw prawn, proffered between
my gloved fingers as bait. The hope was that the scent would waft
into our quarry's home and draw him out, like fresh ground coffee
coaxes the cranky human from their bed.

All of our dives had been spent like this; just lying on the bottom for about an hour until our air ran dry and we were blue with cold. On one dive a mighty wolf eel had come to investigate my prawn. Wolf eels are the largest of the blennies, which are usually the tiddler fish you find in rock pools. These, though, are gargantuan purple speckled beasts, hilariously ugly, with two metres of moray eel-esque body ending in a troll's head. Messy long thin teeth spilled out of its mouth, as it first took the proffered prawn, then proceeded to devour a sea urchin whole in front of me. It looked like a toddler trying to eat a hedgehog.

The next day I was at the surface in a kayak, as killer whales danced about me, spinning, pirouetting and breaching, a huge bull giving me a flyby so close I could have run my hand down his flank, his mighty black dorsal fin towering over my head.

And on the murkiest dive of the trip, when visibility was down to less than two metres of green soup, a group of maybe 20 Steller sea lions came calling. The females playfully tugged on my mask straps and nibbled my fins, before a male that must have weighed three-quarters of a tonne buzzed over my head like a Lancaster bomber, mouth gaping and barking in fury, huge canine teeth bared in an unmistakable threat display.

Every dive we swam through towering kelp forests to get to the den, light cutting down through the fronds in perfect beams. On the seabed, the blooming plumose anemones shone ivory white, like albino bonsai oak trees. It was a Dr Seuss landscape, even more so when you caught sight of a giant nudibranch (a sausage-length bright orange sea slug with a flaming mane around its head) rearing up like a Chinese dragon before plunging its face down into an anemone to devour it.

We dived night and day, but our shy sub-aquatic superstar just wouldn't show.

Prologue

I looked at my gauges. Eighty bar, or a third of my air left, and 15 metres below the surface. I probably had another ten minutes at this depth and then I'd have to surface. It was our last dive, and we'd be heading back without our encounter.* Working the fingers of my left hand down into the cuff of my right, I found the seal of my glove, and rolled it up my wrist, stripping my hand and fingers of their insulating protection. Exposed to the five-degree waters, the cold was instant needles to the fingertips. I'd not be able to keep the glove off for long. I stretched my naked hand out over the pile of crab shells right to the entrance of the den, and waited. After a minute or two, the cold was genuinely excruciating, but I knew I'd soon have my mitts wrapped round a mug of hot chocolate.

Then a wispy strand of orange unfurled out of the dark den like a carpet being unrolled, placing its tip in the centre of my palm. On the pale underside of the arm were white sucker cups, which stuck to my skin. Sensory nerve endings and olfactory cells read my palm by taste and touch. Then another arm unfurled, and another, until my hand was enshrouded in tentacles.† Then they began to pull, with a force that seemed fit to rip my shoulder clean out of its socket, dragging me in towards the den. The temptation was to rip my hand back. The giant Pacific octopus has a parrot-like beak – the only hard part of its body – that can crush straight through the biggest lobster shells, and venomous saliva that can separate their soft and hard tissues. I held my nerve and my ground.

---

* Viewers think it's fake jeopardy when we 'get' the animal on the last day or dive. Truth is, if it really takes that long then you have to make the film all about the waiting, and the fact it's taken so long. Whereas if you find the animal straight away then you can actually get on and make a film about what that animal is like.

† Octopus arms are not scientifically speaking referred to as 'tentacles', but in this case the description sounds right!

She emerged. (I say 'she' not because of biological certainty, or visual cues of sexual dimorphism, but because that's what she seemed to me.) The arms that were so tactile and wispy at their tips were at their base broader than a baseball bat. The domed mantle was perhaps the size of a basketball. The body would one second flow like muslin in the breeze, then be tense as a body-builder's bicep. The next second it would pucker up into crests and fringes. The colour too fluctuated; light orange when she was relaxed, flashes of scarlet when she was unsure. She surveyed me with giant eyes, unsettling because the pupils were horizontal slits, like a viper eye turned on its side.

After the initial power pull, her touch became light, and I led her weightless form out into the open as if taking a partner onto a dance floor. Her touches were undeniably inquisitive, and favoured skin over wetsuit neoprene. The tips of her arms stretched up towards my face, and toyed with the flesh exposed around my neck. We spun together over the seabed in a waltz I'll never forget. A nocturnal hunter exposed in broad daylight, vulnerable to the sea lions and orca that hunt by sight, there was no reason for her to be out in the open. She came out not to feed or to intimidate a foe, but because she wanted to find out about me. An invertebrate driven by curiosity, by something novel in its world.

You have to go back more than 600 million years to find a common ancestor between my kind and hers, a weird worm-like being that looked like neither of us. She was a mollusc; a group that contains slugs, snails and scallops. I am an air-breathing mammal; most of my kin are rats and bats.* Yet even the most militant anti-anthropomorphism animal behaviourist would have to admit ... we had a moment.

---

* Rodents and bats are the most diverse mammal groups, with more species than all the other mammal types combined.

Introduction

The cure for anything is salt water:
sweat, tears or the sea.

Karen Blixen, *Seven Gothic Tales*

As we scour the genomes of obscure species, and scan the heavens for fantastically remote cosmic phenomena, we risk forgetting that the greatest undiscovered country is right before our eyes. The *average* depth of our oceans is around 3,688 metres, and little of it is mapped in any meaningful way. The Ring of Fire, a region of intense vulcanism that encircles the Pacific, is bordered to the west by a canyon that runs from Kamchatka, Russia, all the way down to New Zealand and is 10,000 metres deep.

So much of our oceans is out of sight, and out of mind. We think of them too little, and understand them even less. The deep sea is the largest inhabitable environment in our known universe, and the least understood. It accounts for 95 per cent of the liveable space on our planet. There's an old cliché that we 'know more about the surface of the moon than the deep ocean'. Actually the two are not even comparable; the furthest stars are an open book compared to the darkest depths of our seas.

The oceans seem infinite, too vast for contemplation. It might therefore seem far-fetched that we could fish out an ocean the size of the Pacific. But with 8 billion hungry people, and the unethical, short-sighted and selfish willing to strip-mine our seas, we could empty them completely of life, while filling them with the detritus of a broken society, in my lifetime and yours.

This book is an attempt to script the oceans as if they were a stage play – a revenge tragedy, perhaps. To write backstory and character development for stoic heroes, give insight into the motivation of villainy, before resolution in the final act. And what heroes are here: the tompot blenny and sarcastic fringehead, the weedy sea dragon, the manatee, the mantis shrimp, fretting their hour upon the stage, then gone.

· · ·

The old question of 'what would my 11-year-old self say if he could see me now?' is something my crew and I chuckle about round campfires and ship's galley rum sessions. At 11 I dreamed of being a guide in an African wildlife reserve. I was a fantasist and a dreamer, but even then would have thought my present job ludicrous whimsy. The careers computer at my comprehensive decreed that I should be a fireman. A worthy (and much more useful) livelihood for sure. Imagine if I'd gone round telling my contemporaries, 'No! I'm going to be a shark-whisperer!' I'd have been beaten to a bleating pulp and dumped head down in a bin behind the bike sheds.

My school years were the only unhappy time of a very lucky life. In my sprawling school (located in the midst of a then shabby council estate) the things that made you cool were smoking, fighting, flirting and basketball. I was good at none of them. Instead I was a bright kid who liked writing poems and going for long runs over the Surrey heaths with his rescue dog. I milked the goats on our smallholding before school, knew the location of every badger sett and fox's earth in the local woods, and devoured encyclopedias full of geeky facts about snakes and scorpions. Obviously I could never let anyone at school see the real me, so I made up a tough kid persona, which no one really bought. I was unhappy and lonely.

After a while I just stopped going to class. I'd come in for registration in the mornings and would then 'bunk off', wandering on my own into the woods and heath, looking for birds' nests. My end-of-year report for one subject was simply the sentence 'Who is this boy?' Not surprisingly, I didn't do well in my exams and eventually decided to leave and finish my schooling elsewhere. If it wasn't for one dynamic, inspiring teacher who rocked up in my final year of A levels, my life would have been very different indeed.

I was in my late twenties when I went back to university to study biology and the natural sciences, and was in my forties before I graduated with a master's in bioscience from Canterbury Christ Church University and could finally think of myself as a scientist.

Much as I hated school, my childhood as a whole was very happy. I am blessed with parents who broke their own boundaries, and are in their own way far more adventurous than me. Both come from working-class stock; no one in my family had ever stayed in school beyond 16, let alone gone to university. However, both Mum and Dad saw an opportunity for adventure in working for the airlines and went to work for British Airways fresh from school. They used their discount tickets to travel together, then with my sister and me, to some of the wildest and most exotic places, but on a wafer-thin budget and with an ethos I cannot even imagine attempting with my own kids.

Dad would have liked me to be a rugby player. Mum just wanted me to have a girlfriend. They never pushed too hard, though, and never forced my sister and me to do anything – with the exception of swimming. It was their absolute insistence that Jo and I be water babies, both for our safety and for the benefit of marine-themed holidays. We started learning to swim before we could walk, at the military baths in Aldershot. Stern, shouty sergeants would poke us and our fellow toddlers with long sticks any time we showed signs of coming into the side. Together we choked, spluttered and struggled our way through our bronze, silver and gold survival awards, which were ceremoniously sewn onto our saggy cloth swimming costumes.

The 'survival' part of the qualification was taken very seriously. We were made to dive to the bottom of the pool to collect sunken house bricks, and learned how to inflate our pyjamas into floats. I'm not imagining this. We really took our pyjamas into the pool and blew them up to use them as life jackets. A priceless skill that in 50 years of swimming I've never heard of anyone using. The methods were harsh, and frankly traumatising, but they worked. Jo and I swam our first mile when in primary school. I was six and she was four. Four! She swam a whole mile when she was four! A victory she still lords over me to this day.

With the two of us as happy as fishes in the water, it left Mum and Dad to use their free flights to take us to the most special of seas. My first tropical coral reef was Malindi in Kenya when I was barely seven or eight; I had no idea such eye-poppingly colourful and dazzlingly exotic fish existed. I was hooked.

I certified as a scuba diver, getting a cheap PADI package, while on a year travelling solo around Asia in 1990. It wasn't until I got my dream job with National Geographic in 1998 (with the extravagant title of 'Adventurer in Residence') that I became a professional diver. With a level of luck I struggle to believe even now, my first ever diving series was entitled *Great Dives of the World*, and took me on a round-the-planet trip, doing the finest scuba dives imaginable, from Micronesia to the Caribbean.

In the decades since, I've racked up thousands of hours underwater, and dived with everything from hippos (by accident, not advisable) to six-metre anacondas (on purpose, but also not advisable), under Arctic and Antarctic ice, explored unmapped cave systems and unknown ocean seamounts. Over three decades, diving went from a whimsical dream, to my hobby, to my job.

• • •

I recently celebrated my fiftieth birthday. That significant mile-stone provides me with a personal metric for measuring how our planet, and particularly our oceans, have changed during that time. I don't think of myself as being an old man. Yet the world's population has doubled since I was born. It was 3.9 billion then and is 8 billion now.* Every one of those new people has increased demands on the world's resources, and increased impact on the world around them; and you can see it.

When I was a kid, as a knee-high splasher snorkelling in the Med off Turkey or Yugoslavia, the waters were filled with fish in great colourful shoals. We caught octopuses with our bare hands, and could fish sardines out of the shallows with a hand net. Destructive industrial fishing of the Med only began around the year of my birth, and now the Med's fish populations are teetering on extinction.† In my lifetime, more than 41 per cent of marine mammals and 34 per cent of fish life in the Med have

---

* It was a mere 2.3 billion when my parents were born.
† Ninety-three per cent of fish species are overexploited, targeted beyond scientifically set sustainable limits.

gone. Dive in the Med's clear waters now, and it sometimes feels like a silent blue morgue.

When I started school, summer evening drives down country lanes in my dad's Austin 7 would result in a moth snow – a blizzard of airborne insects that would smear the windscreen and pepper the bonnet with their squished remains. This is a distant memory now, our insect populations having been vaporised by pesticides and industrialised farming practices.

In my lifetime the world's wild animal populations have decreased by 60 per cent. There are half the amount of fish swimming in our seas now than when I was born, and we have conservatively taken 5 billion sharks from the oceans, most so that their fins can end up in soup.

The first transparent plastic bottle hit our shelves when I was a foetus. Since then we've generated 8 billion tonnes of plastic, of which 8 million tonnes a year ends up in the sea. We've lost the Yangtze baiji dolphin, the Caribbean monk seal, the Japanese sea lion, the great auk, the lost shark, the smooth handfish, and the giant Steller's sea cow.* The world's smallest porpoise, the vaquita, will be gone by the time this book is languishing in the remainder bins, and we're on the brink of losing the Irawaddy dolphin and the Borneo shark. And when I say lost, I don't mean we have absent-mindedly misplaced them. Instead, we have hunted these marine species with such short-sighted ferocity that we have eradicated them from existence.

This sixth extinction event we are thundering towards is affecting our seas first. And impacting marine life and ecosystems disproportionately. It's time to take stock. But in the words of the great Jacques-Yves Cousteau, 'People protect what they love,'

---

* I've stood alongside a complete skeleton of one of these giant manatees in Russian Kamchatka, and it appears more like a whale than its living Sirenian relatives.

and to love an environment, first you have to get to know it. I'm not an oceanographer but a biologist, so I'm going to write about our oceans through the lens of its inhabitants, the marine creatures that call them home. And while I may be beginning with words of woe, overall I remain optimistic about the future of our oceans, and also about what our oceans can do to help us to avert the great catastrophe looming over us.

CHAPTER ONE

In the Beginning

The sea is everything. It covers seven tenths of the
terrestrial globe. Its breath is pure and healthy. It is an
immense desert, where man is never lonely, for he feels
life stirring on all sides. The sea is only the embodiment
of a supernatural and wonderful existence. It is nothing
but love and emotion; it is the Living Infinite.

Jules Verne, *Twenty Thousand Leagues Under the Sea*

*The dark dragon slumbers on the black volcanic rock. White
spume drenches him and his fellows, windblown spray falling
like snowflakes. He blinks with disdain then snorts, an explosive
sneeze that the wind blows back into his own face. Atop his head
is a white crust where the snot sneezes have dried out in the equatorial sun, leaving behind a salty crown. Warmed by the sun*

and the dark rock, the marine iguana lifts his body slightly, and swinging it from side to side wanders down into the surf. This is the only kind of lizard that enters the sea to feed, and only here in the Galapagos, the laboratory of evolution in all its oddness.

As the lizard enters the water, so do I, allowing the swell to lift and carry me, trying to become weightless so I bounce over the jagged rocks. Underwater the marine iguana is transformed, and the slumbering shape becomes a whirr of paddling tail as it heads down to the rocks below. Once there, bird of prey talons help him cling onto the tiniest imperfections in the rocks. His short blunt face enables him to nibble even the sparsest algae coverings on the boulders. When he is immersed in his feeding, he is utterly oblivious to me. I push so close that he is practically nibbling my mask.

This is the most primeval of all marine experiences. A living dinosaur* munching on the black volcanic rocks that mark the easternmost rim of the Ring of Fire. And while the destructive force of vulcanism may have been responsible for several of the great extinction events in our planet's history, they can bring life too. In fact, it is possible that the bubbling furnace beneath the waves is where life on our planet first began. The Galapagos is best known as the cauldron of evolution that led to Darwin's dangerous little idea. Less well known is that the discovery of hydrothermal vents here led to a whole new idea of how life might have begun on the only inhabitable planet in the observable universe.

. . .

---

* I mean this only in the aesthetic sense – marine iguanas might look like dinosaurs, but are no actual relation.

You cannot tell the story of life on earth without starting with our oceans. Before life first bloomed, our planet was a truly hellish place. If we rewind back 4.5 billion years to the early Hadean period, rampant vulcanism would have steamed the fluid from the rock of the earth. This fluid would have included galactic ice from comets, and space water held within the foundling rocks of earth. The toxic firmament blazed with clouds, boiling torrents streamed through the lava fields and across shifting tectonic plates. Then, by 4.4 billion years ago, our planet began to cool, and it started to rain. And it didn't stop for several thousand years. This rain formed our oceans. By around 3 billion years ago, almost the entire planet was covered in water.

That this water could persist on earth is down to serendipity. It's often said that our whole planet is a 'Goldilocks zone', not as near the sun as the furiously hot Venus where the surface temperature would melt metal, or as far as the chilly Mars where water is present only as ice. Our planet, like that fairytale porridge, is neither too hot nor too cold.

When I first studied biology at school, the dominant hypothesis for the creation of the first life on earth was still the 'warm little pond' suggested by Darwin. In 1953, scientists at the University of Chicago subjected a soup of chemicals to what they saw as the conditions at the dawn of time. They used sunlight and pseudo lightning with an 'ocean' of liquid water and a gaseous atmosphere, and eventually managed to create amino acids – one of the building blocks for life.

This Miller–Urey experiment has now been bypassed by several other hypotheses. Evidence of 'building block' amino acids within meteorites led to the 'panspermia' theory, in which it is posited that the microbes that sparked first life arrived on projectiles from space. As our planet is only a third of the age of our universe, life could have been cooking away in some other galaxy far, far away for billions of years.*

Another idea is that life began at what is now our planet's most inhospitable ecosystem – the hydrothermal vent. Discovered in the Galapagos in 1977, these fissures in the deep spew out super-heated water and toxic chemicals, at a depth where the pressure would crush a watermelon to the size of a raisin. The water rushes upwards in plumes, the colour and heat of which defines them as either 'black smokers' or 'white smokers'. Despite these living infernos being so inhospitable, these vents are actually oases in the deserts of the abyssal plains. Almost every research dive to investigate a hydrothermal vent uncovers bizarre new species of life: scaly-foot gastropods, limpets, giant vent mussels, snails, eel-like fish. Or Pompeii worms, whose snouts sit in the 50°C water of the vents but whose tails freeze out in the ocean cold at –2°C (like if your feet were in an ice bath, and your head was in a pizza oven).

Microbes and bacteria that feed off chemicals like hydrogen sulphide and methane live here at the edge of the vents. These organisms are known as 'chemotrophs' (chemical consumers). Instead of photosynthesis, they grow through 'chemosynthesis', in which chemical energy is used to transfer carbon dioxide into sugars in the complete absence of sunlight. As these conditions and chemicals would have prevailed in the infant earth and have

---

* Indeed cosmologists developed their theory of how our oceans formed by studying the isotopes of elements in the earth and other icy bodies in the solar system. Our planet first came into being during what was known as the late heavy bombardment period, with heavy elements colliding in space and coming together into a spinning ball. This was when our moon was formed, probably smashed into orbit as ejecta from a head-on collision with our planet. (This is known as the giant impact hypothesis.) During the cosmic bombardment of our foundling planet, icy comets and planetesimals crashed into the rocky surface, depositing space water. Simply put, this space water is the origin of all water on earth today.

persisted unchanged for billions of years, it seems the inferno might have been the perfect place for genesis to commence.

Black smokers today are inhabited by microscopic entities with different biochemistry to other microbes. These thermophilic (heat-loving) extremophiles (extreme-lovers) appear to be the most primitive of organisms, thought to have formed 4 billion years ago in the Hadean era, and named after the fires of Hades itself.* These are the archaea; single-celled organisms that we have in our guts, mouths and skin.

The weird animals around the vents live in another 'Goldilocks zone', where they can either consume microbes, or suck them in and steal their energy. These include the hairy-clawed yeti crab, grotesque swarms of blind shrimp, and giant tube worms who have no gut or mouth, but devour oxygen, carbon dioxide and hydrogen sulphide through a feather-like red plume, perfuse with blood vessels. These extremophile oddities include 'the Hoff' squat lobster, so named because it sports a luxurious hairy chest, just like its *Baywatch* namesake.

Everything here at the boiling vent edge seems impossible. And as a very lapsed Catholic, I find it a gloriously heretical idea that Genesis did not happen in the heavens, but that life might have instead been born in the hellfire of Hades.

• • •

It was perhaps 3.8 billion years ago that the first single-celled organism appeared in the sea. About 2.4 to 2 billion years ago blue green algae or cyanobacteria started to photosynthesise, creating oxygen as a byproduct, which led to our atmosphere starting to become oxygen-rich. To us this cyanobacteria would appear as nothing grander than slime, but it was the building

* The zone of the super deep below 6,000 metres is also known as the hadal zone for the same reason, honouring the Greek god of the underworld.

block which paved the way for all oxygen-consuming life that came after. This is known as the Great Oxygenation Event. For much of the life that had till this point developed in an anaerobic biosphere, this was a catastrophe, as they now found the oxygen-rich atmosphere to be toxic.

Around 650 million years ago the first supercontinent, Rodinia, began to break apart. This would have coincided with the great glaciation event known as Snowball Earth. At this period in time (the Cryogenian) the whole surface of the planet may have been covered in a giant ice sheet. There would have been seasonal melting, leading to periods of slushy or even open ocean. This ended by around 635 million years ago, when animals with soft, sponge-like bodies in a wide variety of shapes, including discs and tubes, emerged during the Ediacaran period, a time of extraordinary diversification. A few tube-like creatures might also have had firm outer coverings or shells. Down on the seabed a carpet of microbes began to form, and symmetrical oval oddities like *Dickinsonia* cropped up to feed on them. The fossil evidence of the Ediacaran is slim, and mostly limited to famous deposits in the Australian hills, but for the next 15 million years, they were the apex of evolution. Even at this point, life still only existed in the deep oceans.

The first evidence of plant life on land came shortly before the Cambrian explosion (around 541 million years ago). The ancestors of all species alive today started to appear in the fossil record from this time. Animals began to diversify, and to become more complex and sophisticated.* Cockroach-like trilobites abounded, with perhaps 20,000 species, as did spiny, slug-like jellyfish, anemones, sponges and arthropods with jointed legs. In

---

* It is important, however, to not fall into the trap of believing that evolution is a never-ending 'search' for complexity, with us and our descendants as the ultimate goal.

2019 a stunning discovery in China revealed 20,000 fossils from Cambrian seas, so well preserved that gills, soft tissues, mouths, eyes and guts could be identified. The largest predator of the Cambrian was *Anomalocaris*. At around a metre in length, with bulging eyes and bizarre bodies, they used curly mouthparts to draw prey in towards slicing teeth. Scales found from this era might well have belonged to the very first shark ancestors, though we know little about what they might have looked like.

Most of this life was eradicated by the first of the cataclysmic mass extinction events, around 440 million years ago. This was the Ordovician mass extinction, and, like all other extinction events before or since, it hit life in our seas first and worst. Perhaps 85 per cent of life was eradicated, almost certainly as a result of climate change.

Eventually, though, life in the oceans started to recover. Around 335 million years ago our landmasses coalesced into a supercontinent known as Pangea, surrounded by the all-conquering ocean Panthalassa, potentially the greatest living ecosystem ever known. Within this vastness hunted two-metre-long sea scorpions called eurypterids, and huge numbers of cephalopods. Jawless fish called ostracoderms gave way to placoderms like *Dunkleosteus*. This predator was the size of a modern great white shark,* but was clad in armour plates, and instead of teeth its jaws were self-sharpening bones that it used to shear through its prey. Swaying fields of crinoids or sea lilies covered huge areas of the seabed. Today these remain, mostly as inhabitants of the deepest seas.

In the Permian, 299–251 million years ago, we see some of the most bizarre of the proto-sharks. Shoals of metre-long *Stethacanthus* hunted with what looked a bit like a shower head

---

* Recent research disappointingly suggests it might have been much smaller.

stuck to their backs, an appendage that might have been a sucker to allow them to attach themselves to bigger creatures, much like a modern-day remora does. There were giants like *Helicoprion*, which in body is similar to a modern shark but appears to have swallowed a buzz saw. The disc whorl that sits vertically in its jaws is well represented in the fossil record. Some are the size of dustbin lids, so *Helicoprion* could have been as long as a bus.

Much of what we know about the past tends to be defined by catastrophic extinction events that eradicated the majority of life on earth. The end-Permian extinction that occurred around 251.9 million years ago is the biggest, with an estimated 90 per cent of marine life disappearing. Life eased away with a whimper rather than a bang in a slow extinction event that took around 60,000 years. Again it appears that climate change was a factor, probably caused by a highly volatile period of volcanic activity, with the Siberian Traps volcanoes all erupting simultaneously. Coral reefs were brutally hit and took millions of years to recover.

The supercontinent split up around this time, and began moving again. The Pacific was formed around 250 million years ago, and then the Atlantic. This heralded the best-known time for prehistoric life on earth: the age of the dinosaurs and their cousins in the sea, the marine reptiles. High sea levels and generally warm global temperatures throughout the Jurassic and Cretaceous led to the formation of large areas of warm shallow seas, where marine life flourished. Toothed fish, sharks, reptiles, birds and pterosaurs plunged into these warm waters to find food.

This next hundred million years is arguably the most exciting for our oceans. If I could travel back in time, this is when I would visit; though I would take a heavily armoured submarine. The oceans were dominated by giant reptiles that fed on fish, bivalves and each other, including ichthyosaurs, which are often described as dolphin dinosaurs (despite actually being neither).

One ichthyosaur, *Ophthalmosaurus*, had the largest eyes and probably the finest visual acuity of any predator that has ever lived. Plesiosaurs may have evolved from terrestrial predators, with their classic 'Nessie' necks and small heads, some of which might have ploughed through seabed sands to snatch up buried sharks. Relative latecomers to the party, but perhaps most frightening of all, the mosasaurs were pure meat-eating machines the size of whales. Giant turtles like *Archelon* were around by the Cretaceous 146 million years ago and looked just like modern turtles – apart from being four and a half metres long – and might well have been munched by some of these other swimming horror shows.

Around 170 million years ago there was a 'regime change' in the seas.* Until that point non-biological factors such as ocean chemistry and climate had been the biggest forces driving evolutionary change, but now it was the proliferation of plankton, which formed their own calcium carbonate skeletons, that changed the order of things. These stabilised life and balanced out the acidity in our seas. Ecosystems were moderated, and organisms were less susceptible to climate change and corresponding ocean acidification than before. Now predator–prey interactions became the most important factor in driving the evolution of life. Creatures evolved defences; spines, spikes, venoms and poisons, thick shells, and armour to stave off attackers. The predators developed their own weapons, becoming faster, more wily, and armed to the teeth. The era of the evolutionary arms race was here.

The most infamous of all extinction events was approximately 65 million years ago.† A six-mile-wide asteroid smashed into Chicxulub, at what is now the Yucatan peninsula in Mexico.

---

* Based on recent research by the University of Plymouth.
† Marking the end of the Cretaceous period and beginning of the Tertiary, the so-called K-T Boundary.

Had the asteroid plummeted into our biggest habitat of the deep sea, then tsunamis would have obliterated everything near to the shore, and caused dramatic but short-lived devastation. Instead it smashed into shallow seas, and blasted the minerals of the seabed up into the atmosphere, causing dust clouds that circled the globe and blocked out the sun. This calamitous event led to tsunamis, wildfires, acid rain and rapid global cooling.* Over the course of decades 75 per cent of life disappeared, including all of the non-avian dinosaurs.† And yet life persisted, both on land and in the seas. In fact, as the dinosaurs and marine reptiles disappeared, it paved the way for the birds, mammals and fishes to leap forward.

The first marine creatures that are recognisably whale-like date from about 50 million years ago. About 34 million years ago the temperatures in the oceans fell due to shifts in the tectonic plates. Then 30 million years ago Antarctica and South America drifted apart, and the Southern Ocean was formed. As North and South America came together, the isthmus of Panama was formed, and the Pacific and Atlantic oceans were cut off from one another. A warming Caribbean gave rise to the Gulf Stream, which remains perhaps the most important force in oceanography and meteorology to this day. Ocean giants such as the blue whale are evolutionary newborns, only really developing around 5 million years ago, perhaps to potentially end up as a meal for the legendary shark megalodon.

From 115,000 until 11,000 years ago much of the planet was subjected to an Ice Age, with the majority of the United Kingdom covered in an ice sheet. With so much water bound up in ice, sea levels were radically lower. You could have walked from

---

* It might have been supplemented by other factors such as eruptions at the Deccan Traps in what is now India.
† Though not, intriguingly, the crocodilians and turtles.

England to mainland Europe on the dry plains of Doggerland, even as agriculture was starting up across the continent.

Spine-tingling proof of this can be found when diving on the seabed in the eastern English Channel. I've swum through gloopy green water 20 metres below the surface, past the roots of a fossilised oak forest, over the crude planks that my Ice Age ancestors gathered to make docks for their coracles and the burnt ash of their fireplaces. In one memorable place, a lobster had excavated a burrow for itself, and dug out a spoil pile of chipped flints from the mud. Taking some of them in my hand, I could clearly see these were worked flint tools, chipped into razor-sharp blades for scraping hides and slicing meat. No human had handled those crude knives for 10,000 years.

. . .

In the 1970s James Lovelock suggested the Gaia hypothesis (named after the primordial Greek goddess of the earth). Gaia states that living organisms interact with the environment to form an infinitely complex and self-sustaining system. The maintenance of the hydrosphere* was a central part of this theory. The water that arrived on cosmic comets and was steamed from the rocks is what fills our aquifers, packs our lakes and glaciers, and trickles down our mountain streams even today.† The state and location of this water is in constant flux – although the world cools and warms, becoming more lush, frozen in ice ages or increasingly arid, the mass of water in our water system remains constant.‡ It might lie sleeping in million-year-old Antarctic ice,

---

* All the water existing within the earth's terrestrial systems and atmosphere.
† An alternative hypothesis states that the interplanetary collision with the planet Theia that led to the formation of our moon also led to the creation of most of the earth's water. These theses are not mutually exclusive.
‡ With the exception of microscopic amounts of hydrogen lost through the upper atmosphere into space. This effect might have been far greater in the young earth.

form as clouds, fall as rain, seep through the crust and be boiled back to the surface to continue the cycle, but the whole does not increase or decrease.*

Of the water available to us today, 97 per cent is bound up in our oceans. We are only just now starting to wake up to the fact that the water we have is more precious than gems or fossil fuels could ever be. Every drop we consume came from our primordial rock and the heavens, and may have passed through the digestive tract of half a dozen extinct species. If we need any extra imperative not to sully our water, surely this is it.

---

* Potentially minute amounts could escape through the upper atmosphere, but this may well be balanced against water arriving on meteorites.

CHAPTER TWO

Abyss

With its untold depths, couldn't the sea keep alive such
huge specimens of life from another age, this sea that
never changes while the land masses undergo almost
continuous alteration? Couldn't the heart of the ocean
hide the last-remaining varieties of these titanic species,
for whom years are centuries and centuries millennia?

Jules Verne, *Twenty Thousand Leagues Under the Sea*

*With a gentle bump, our sub settles on a windowsill ledge, the dull
bell chime of thick metal on rock vibrating through our bones like a
tuning fork. For a minute or so, we sit in silence and darkness, as
the clouds of slow silt settle, then pilot Gary flicks on the headlights,
illuminating a new world.*

*For a second I am the first astronaut looking through the port-
hole at the surface of the moon. All that is needed to complete the*

21

picture is a line of Buzz Aldrin's footprints across the grey. The silt bottom we sit on looks loose and floaty, a towering rockface dusted with ivory powder looms over us, and to one side the wall drops off into the abyss. Beyond the beams of the lamps is thick darkness, and fine marine snow glitters like diamond dust in our light.

There is hardly a square inch of bare rock. Everywhere is covered in gorgonians or soft corals, and of course sponges in every colour imaginable: bright oranges, reds, yellows, pinks and greens, to name but a few. Aside from a Pimms-like lake full of flamingos and budgerigars, I can't imagine a more impressive explosion of colour. There are giant elephant ear sponges, barrel sponges, and tube and vase sponges, ranging from bright white to pure black, all attached to the sea wall where the sea snow doesn't block their pores. Sea whips and rope sponges seem about to lasso us and drag us into a wall that boils with life. There's nothing in nature, not that I've witnessed, that even comes close to this. It's the most beautiful thing I've ever seen.

• • •

It was June 1930 when explorer William Beebe and engineer Otis Barton took the first look at our deep sea. They travelled down in a circular steel bell, with a single smeary window of quartz and an open tub of lime inside to soak up the carbon dioxide they breathed out. The intrepid pair bottomed out at an extraordinary depth of 1,426 feet (approximately 435 metres), reporting back with the morbid and poetic epithet that 'only the dead have been deeper than this'. As their expeditions became more and more bold, they pushed deeper still, bringing back tales of animals that glowed in the dark, pulsating jellyfish and odd marine worms, the original gutless wonders.*

---

* Jellyfish have no intestine, liver or pancreas to help in digestion; instead, they have a simple digestive cavity, and both take in and excrete food through the same orifice. Though many worms do have the equivalent of a gut, deep-sea marine worms such as *Osedax*, the blind zombie worm, do not.

If you dropped Beebe's steel bell into the water untethered over our deepest abyss, it would take at least half an hour to sink to the bottom. In its long journey it would pass through five main zones.

The photic, sunlight or epipelagic zone is the area where light is able to penetrate and is generally considered to be from the surface down to about 200 metres in depth. Heat from the sun is responsible for the wide range of temperatures that occur in this zone. Within this part of our ocean plankton can photosynthesise, and dolphins, seals and porpoise play and hunt. It is the most productive and busiest layer of the ocean.

From the bottom of the epipelagic down to 1,000 metres in depth is the mesopelagic, also known as the twilight or mid-water zone. There is still very faint light here, and the first flashes of bioluminescence from some animals. It's still laden with life, with about 90 per cent of the world's fish occurring here (in terms of biomass, not biodiversity), as well as squid, krill and jellies. The tiny bristlemouth fish (named for its symmetrical bristle-like teeth) alone may number into the quadrillions (I did not make that number up!), making them the most numerous vertebrates on earth. Some of these creatures (especially zooplankton) practise what's known as diel vertical migrations, travelling up to the surface at night, then retreating at daybreak to avoid their predators. This represents the largest daily movement of life on our planet.

About three-quarters of the world's water is *below* this depth. The bathypelagic or midnight zone stretches from 1,000 metres down to 4,000 metres. As I mentioned earlier, this is, unbelievably, the *average* depth of the seabed, and includes the abyssal plain, an area of seabed with relatively level stretches, making up 70 per cent of the sea floor. Possibly the weirdest of all ocean environments are found here: brine lakes made up of incredibly salty water on the seabed. These are the remnants of ancient seas that existed tens of millions of years ago. Animals that fall into these lakes are generally pickled alive. However, living near them can be valuable, as they are rich in methane, and some organisms can use this to create chemical reactions and energy.

The blooms of life at the abyssal plain are around hydro-thermal vents, or 'whale falls', where a giant whale will die and sink slowly to the bottom, carrying its nutrients with it. The hagfish and sleeper sharks of the midnight zone may take years to strip a large whale to the bare bones, and then special-ised feeders like the charmingly named flame-haired 'zombie worms' turn up to finish off the skeleton. Some whale falls can support a worm density of 45,000 per square metre, the highest concentration of life in our oceans. Though they may be comparatively rare, these whale falls number perhaps half a million worldwide and provide stepping stones along migratory routes, like oases that enable certain organisms to hop from one busy deep-sea hot spot to another.

The only visible light at this depth is from animals that produce it themselves. Most animals are black or red due to the absence of light. They have squishy and slimy skin and equal pressure within their bodies than without, allowing them to withstand pressures 110 times that found on the surface. These creatures that survive on little food may have super-slow meta-bolisms, move slowly, mature late, and live for a long time.

Only 5 per cent of species ever travel to and from these depths and the photic zone. However, the deepest-diving mammals *can* get down this far. The Antarctic Weddell seal has been one of the most comprehensively studied in their plummets to depth. They can make it to 1,000 metres and stay submerged for 70 minutes. To achieve this, they store twice as much oxygen as humans can in their haemoglobin

and myoglobin. While a human freediver may keep 36 per cent
of stored air in the lungs, 13 per cent in the muscles and 51 per
cent in the blood, Weddells keep only 7 per cent in their lungs,
28 per cent in the muscles and 65 per cent in their blood.
They have double the blood volume that we humans do; even
more if you consider that so much is also stored in the blubber.
They have an unparalleled ability to control both their heart
rate and the dilation of their venous system. Most mammals
have 13 sets of ribs. Weddell seals have 15, to allow for larger
lung space. Their muscles are capable of storing ten times more
oxygen than a human of the same size, and are so deeply red
that they appear black.

Elephant seals can also penetrate to 1,000 metres, and
Cuvier's beaked whales have been recorded diving for three
hours and forty-two minutes and to a depth of 2,990 metres,
but it's likely the male sperm whale can go deeper still. They
have been timed diving for over two hours, after which they have
returned with their stomachs filled with bottom-living sharks in
seas over two miles deep.*

While working with scientists studying sperm whales diving
into the deep-sea canyon off Kaikoura in New Zealand, I was
allowed to sit silently in my kayak alongside sperm whales at
the surface as they prepared for their dives. You soon learn the
patterns of the whales so precisely that you could set your watch
by them. For the first few minutes after they surface, they heave
out great puffs of dank, fish-smelling air. Towards the end of
an approximately eight-minute breathing cycle at the surface
– during which time they barely move – the whale takes some
huge breaths, powers itself forward at the surface and then dives.
However, before they drop, they breathe out; the air is not being

---

* Sadly, much of what we know of basic cetacean biology comes from the
records of whaling vessels that have caught and killed the animals.

kept in their lungs (which compress to a tiny size at great depth) but in their muscles and blood.

Sperm whales' blood makes up 20 per cent of their weight, packed with red blood cells running through arteries so wide it's said a child could crawl through them. Their blood is as thick as gravy. Oxygen is stored both in haemoglobin in the blood and myoglobin in the muscles. When they dive their heart rates drop to no more than five beats a minute, and blood is shunted away from the extremities to essential organs, and to the brain and diving muscles. All of this serves to reduce the use of oxygen in the body, and allow the animals longer in the depths. Using modern tracking tags it's clear that after the initial effort to leave the surface, sperm whales drop rapidly and without effort, which suggests their bodies must overall have a similar density to water.

Despite their final exhalation, studies of sperm whale bones show that most suffer chronic bends* consistently throughout their life.

Where the sperm and Cuvier's beaked whales turn and head to the surface is unbelievably less than halfway to the very bottom. The next zone is the abyssopelagic – the appropriately named abyss – which descends to 6,000 metres below the surface. The word comes from the Greek 'without bottom'. The temperature is constantly near freezing even in the tropics.

All the way down to 6,000 metres there are still coral reefs. Like those we see in shallow tropical seas, these can be found as colonies of different species, or as individual coral polyps, and are a habitat for thousands of creatures. However, rather than

---

* The bends is a diving injury where bubbles of nitrogen get into the tissues – especially the joints – of a diver, or animal, at depth. When returning to the surface without proper decompression, the bubbles expand, which can cause crippling injuries in humans and even death.

gaining energy through photosynthesis, the corals trap passing nutrients from the water column.

The deepest ever fish discovered was a cusk eel trawled from 8,372 metres in the Puerto Rican trench. Below this, it seems there is only invertebrate life.* This is the hadalpelagic zone, the very bottom of the deepest ocean trenches. The Challenger Deep, a narrow slot within the already precipitous Mariana Trench, running in an arc between Japan, the Philippines and New Guinea, is the deepest of these trenches at 10,900 metres. Some organic matter can take years to drift its way down to its depths.

The first descent to the very bottom of the Challenger Deep was made in 1960, in the bathyscaphe *Trieste*, piloted by Captain Jacques Piccard and his pilot Don Walsh. The next manned descent was by film producer James Cameron in 2012, and in 2019 the DSV *Limiting Factor* began a new era of deep exploration, with untold mysteries and wonders still to be uncovered in the deepest, darkest seas, and even more clues as to the origins of life on earth to be found.

• • •

As I probably won't have the opportunity to explore the deepest depths of our oceans, the closest I am ever likely to get to some of the animals that live there is either when they are bought up onto the deck of a research vessel or in the pickling jars of an institution of learning. One of the finest assemblages in the world is the 'spirit collection'† at the Natural History Museum in London. While what is on display is legendary, what is behind

---

* Although the voyage of the *Trieste* originally claimed to have seen a fish on the bottom of the Challenger Deep, this has been ... challenged!
† So named because the behind-the-scenes collection is all preserved in alcohol.

the scenes is even more dramatic. There are more than 20 million specimens on the 16 miles of storage shelving, making this the most important natural history collection on the planet.

On my first visit, curator Oliver Crimmen led me down several flights of stairs and through three different security doors, all of which had to be beeped open with his digital ID. We passed rows of blue specimen cabinets, all on rollers and shifted side to side with a sort of ship's helm. As the cabinets slid open they revealed shelves loaded with oddities, some of which date back to Darwin's first expeditions.* There was a strong smell of mothballs (all museums suffer from a variety of invertebrate pests nibbling their way through the collections), and of the alcohol and formalin used as embalming fluid. It was a heady aroma that smelled like death when I first went behind the scenes at a natural history museum. Now it smells like knowledge.

Finally, Oliver led me into the marine specimens room – also known as the tank room – deep in the bowels of the building. To many it would seem unspeakably macabre, with countless vials, tubes and glass cases, all filled with preserved animals. While some vessels merely contain organs, heads or limbs, most contain the complete animal. New species have been discovered here, while others no longer exist in the wild. The collection also contains many of the holotypes of certain species – that is to say the type specimen, the single physical collected animal first used to describe their kind to science.

In 2004 the museum was offered a nearly complete giant squid, which had been accidentally caught by a fishing trawler at just 220 metres below the surface. Archie (for *Architeuthis*, its scientific name) was to become the marine collection's prize.

---

* Darwin's common octopus is here, collected in the Cape Verde islands in 1832, taken live on the voyages of the *Beagle*. There is also James Cook's honeyeater, discovered in Hawaii in 1778.

A bespoke nine-metre-long tank was constructed so that Archie could be kept not only intact, but stretched out so that all of its soft form can be appreciated. The only hard part is the beak, which looks like that of a giant macaw, and is kept in a separate jar. Holding this dead and dismembered beak in my hand might be the closest I will ever get to a giant squid.

There were many other icons of natural history there that I knew I wouldn't ever see in the wild. The deep-sea anglerfish was by far the strangest. As one of their treasured specimens, and arguably the most perfectly preserved one on the planet, I was gobsmacked that Oliver allowed me to take it out of the jar and hold it in my hands. This one was a female, box-like in shape, like a heavily deflated old leather football. The skin was dark brown and bumpy. Its giant maw was lined with tiny needle-like teeth, and the minuscule rheumy eyes were recessed and lifeless. In the centre of the head was a flexible fishing pole, which lured down over the cavernous mouth. This is actually a modified dorsal spine, and on the end is a small blob called an esca, within which are light-producing microbes. This gives the esca the ability to bioflouresce, turning it into a lure, attracting prey to swim right into the mouth of this ominous-looking predator. In some anglerfish species, this lure can be longer than the whole animal.

The weirdest thing about this oddball species is its method of reproduction. Swimming around in the barren black of the deep sea, it could be years before one encounters another of its own kind. That being the case, the male has a sneaky – one might say lazy – method of ensuring paternity. The male follows the scent of a female, and when he finally finds one, he attaches to her flanks, after which most of his body and organs are simply re-absorbed. He becomes a pair of gonads, attached to the female, providing her with a constant source of sperm and feeding off her living flesh. Oliver handed me a male. While I needed both hands to hold the weight of the female, the male was little more than the size of my little fingernail and was kept in a plastic case the size of a matchbox.

Oliver next took me to a row of thinner glass cylinders, most about as wide and long as my leg. Inside were sawfish with rostrums lined with teeth, rolled-up rays and bizarre chimera with bulbous bobbly noses. One contained a dwarf lantern-shark (the smallest species of shark), which is fully grown at the length of a standard ruler. Not much bigger but far more feared is the parasitic cookiecutter shark. This cigar-shaped fiend spends its days maybe two miles below the surface but will swim up into shallow waters at night, and it has nightmare gnashers. The top row are pointy and latch on to a food source such as a whale, dolphin or seal. The bottom row is one continuous line of symmetrical sawblade teeth (the whole set is shed at one time). They are thought to hunt in packs, with the fish zipping in and out to their prey and digging in their terrible teeth before spinning around on a horizontal axis and removing a neat circle of flesh – hence the name cookiecutter. Often the most high-value tuna will turn up in market with these perfect and agonising-looking wounds on their silvery flanks. There have also been several recorded attacks on humans. One was a 61-year-old long-distance swimmer attempting to swim by night from Hawaii to Maui. He recalls being bumped several times by small but persistent fish trying to get at him, before one bit him in the chest. As he swam to his support vessel to get out of the water, one latched on to his calf, and took a perfect, neat circular incision out in one bite. It took nine months of skin grafts for the bite to heal. However, bearing in mind the great whites, tigers and bulls that inhabit these waters and are active by night, you might think he got off lightly.

The primary reason for my behind-the-scenes visit, though, was to see the giant metal tanks that give the tank room its name. Inside them are some of the most impressive and most valuable of all the museum's deep-sea specimens. The brushed metal lids on the cavernous containers needed to be lifted with a system of wheels, chains and pulleys, revealing the tannic brown formaldehyde and grotesque pickled monsters lurking inside. For a moment it felt like being in Dr Frankenstein's laboratory. Oliver rolled the wheels, chains clanked, and the lid lifted to reveal one of the strangest assemblages I'd ever seen.

At one end of the tank was a frilled shark. Often considered a living fossil or evolutionary throwback, its long cylindrical anguilliform (eel-like) body doesn't look anything like what most consider a shark to be, although it's possible that many primitive sharks might have looked more like this. This mile-deep monster's common name comes from the weird frills that seem to pour out of its gill slits. Inside its maw are 300 needle-like recurved teeth (curving inwards), bunched together in groups to snag its squid prey.

We pushed aside a porbeagle and a blue shark, both staring lifeless from the brown fluid, to find another denizen of the deep. This was probably the most extreme of all vertebrates and the biggest deep-sea fish yet discovered: the Greenland shark. They are found deep under Arctic ice and live the slowest of lives. This perfect specimen was swept ashore in Embleton Bay in Northumberland, England in October 2013. It's just one peculiarity of Greenland sharks that they are considered to be part of the UK's native fauna, but as they live so deep below the surface here they are almost never encountered.*

---

* Bizarrely, in March 2022 my colleague Professor Rosie Woodroffe found another perfect Greenland shark washed up on Newlyn Beach, just a stone's throw from my wife's family home!

To lift the three-metre-long shark out of its golden-brown liquid morgue, I had to plunge my arms in up to the elbow. My rubber gloves were not quite long enough, and a mix of alcohol and aged shark juice flooded in and soaked my best jumper. Even through the rubber gloves, the coarse sandpaper skin was tangible beneath my fingertips. The teeth had certain similarities to the cookiecutter's – just one of the reasons why the shark is believed to feed mostly on the carcasses of things like whales, sawing its body from side to side to remove chunks of dead flesh. Despite this particular specimen being so perfectly preserved, scientists learned precious little about these sharks' diet from it. Its stomach was empty save for a few tapeworms, which suggests that it probably everted its stomach before it died. The animal was clearly not well. It also did not exhibit the Greenland shark's famous parasitic accomplice – a strange copepod that attaches itself to the animals' eyeballs and feeds off them.*

The Greenland shark hit the headlines in 2016 when research conducted on the eyeballs of the shark garnered some remarkable results. Scientists used radiocarbon dating on the lenses, and the results showed that these sharks can live at least 272 years and possibly more than 500. The females do not reach maturity until they are 100.

These jaw-dropping results told us something important about life in the deep seas. In our coldest and deepest seas, life slows right down. A small deep-sea octopus was observed sitting over her eggs for four and a half years. Most octopus species don't live for more than a year. Any animal that takes this kind of time to regenerate, and is not especially common, is incredibly

---

* There's even a hypothesis that there is a symbiosis here, with the copepod working a little bit like the anglerfish's lure, to attract prey towards the shark's mouth.

vulnerable; you only have to take a handful of them out before the entire species is at risk. There are so many lessons to be learned from this, but we refuse to heed them. Our deep seas are out of sight and out of mind. A good deal of this must surely be how little interaction we have with this forgotten environment. There is no other habitat on earth that receives so little attention in the media, and so few visits from adventurers and scientists. I have spent my whole adult life chasing marine dreams, but have only had one foray into the depths.

• • •

The Bloody Bay Wall starts in shallow waters, both close to and parallel to the northern shore of Little Cayman. For several days prior to my dive, I'd been in a whirlwind of tax haven tourism: whisk-you-in-and-out conveyor-belt dive schools, sweaty, paunchy Americans in lurid beach fashions, surgically enhanced wives in giant hats and Chanel sunglasses. But we were here to journey to another world, as far away from the (to me) uncomfortable flamboyance of conspicuous wealth as ... well, the stars.

The Atlantis deep submersible that was going to convey us beneath the surface was the archetypal yellow submarine, about as long as a transit van but cylindrical and with a round turret about a third of the way down. One end of the cigar shape was a domed window, made of bulletproof glass as thick as the complete works of Shakespeare, the curve giving a very strange perspective to everything outside. Inside the sub, there was space for me and my camera operator to cram up next to each other in front of the glass. Gary the pilot sat behind us steering.

Once the circular hatch was slammed into place, a few wheels were spun and levers pulled. There was a hissing, a bubbling, and then we sank at a surprisingly rapid pace. There was a disorientating sense of vertigo, although we had Bloody Bay's decorous wall in front of us for reference. This is sometimes referred to as the most spectacular wall in the Caribbean. We could have been falling alongside the cliffs of the Grand Canyon, except here the rockfaces were emblazoned with hard and soft corals, nurse sharks and green turtles coursing around angelfish and sinuous marbled moray eels.

We were a hundred metres down and falling out of the light before we'd even really got used to the hatch being closed. Every few seconds I would leap off my hard bench and yell, 'Wow, look, there's a spiny lobster! Jesus, look at the size of that sponge! My God, is that coral?' Though some of the forms were familiar, the unfamiliar experience of seeing them so deep, in the dark and from the window of the submersible, had me leaping up and down in excitement. Poor Gary had to fight to keep up with me – no sooner had he started explaining one phenomenon than I was exploding with excitement about something else.

The descent was at this point only vaguely controlled sinking, and before we knew it, the glitz of the Caymans was a distant memory. There was a bump, and our module had landed, settling its struts in the dust of another world. The scene seemed eerily lunar-like, devoid of life, until my gaze was drawn to the silt below. It was criss-crossed with the trails of organisms living in or on the sand. If you followed a track to its end, a small partially buried form might flutter slightly, revealing a reddish shape – a hermit crab sitting no more than a few metres from the sub. As we squinted, we become aware of more and more life. In the water column before us were transparent, dime-sized, jellyfish-like discs, with backwards tracing spines or feelers with rainbow illuminations pulsating down their length – tiny dinoflagellates like strands of DNA whipped and curled. And then two black jacks – squat, fast fish of a genus regularly seen on the reefs above – flitted past the dome, looking in with their curious dark eyes at the strange aliens inside the yellow whale.

In the silence, the only sound was our breathing. I imagined the soundscape outside to be like that of a howling desert gale. Seemingly so close we could touch it, a tiny toy submarine also settled in the silt. But as our eyes adjusted, we realised it was the other sub that was also part of the fleet, the same size as our own but far away; we were seeing it through water of such clarity that it was hard to even tell it was water. The distortion of our six centimetres of bulletproof glass added to the illusion.

A tubeworm shaped like an anemone swayed on an imperceptible current, catching nutrients off the breeze with its sticky tentacles and stuffing them into its central mouth. Seemingly static crinoids – triffid-esque animals in colonies with roots that grip the rock base – actually shuffled across the seabed. Each had a single long stalk, topped off with a furry, floppy daisy-head crown a bit like a Cossack hat. They had swayed in these depths for 500 million years.

At this depth of around 1,000 feet (300 metres), organisms can still make a living from plucking nutrients from the water column. Much deeper, and the search for food becomes significantly harder. This is the realm of the gutless wonder, including oddities like the blind zombie worm, whose scientific name (*Osedax mucofloris*) translates to 'bone-eating snot flower'. A stalk that protrudes from its mucus-like body extends into the water column to absorb oxygen. It has no digestive system, mouth or anus, instead drilling into and dissolving food like whale bones with tendrils on the underside of its slimy body.

'It's exactly like being in a spaceship,' I murmured banally, finding it impossible to sum up any part of this experience in words.

'Yup, a small spaceship.' To be fair, my pilot Gary didn't exactly outdo me in creative description.

'Would you like me to teach you how to drive?' Gary asked.

'You haven't seen me drive a car!' I quipped. I was guessing that the smallest ding to the multi-million-dollar sub would seriously harm his no-claims discount.

'It's quite a lot easier,' Gary responded. 'This red valve here is how I bring you to the surface if there's an emergency. One thing about being in a submersible is that you can't get lost, because home is straight up.'

'Believe me, I could get lost.'

Gary pointed over into the abyss. 'It's 6,000 feet to the bottom there,' he said. That was disappointing. I kind of thought we were on it already, but instead we were resting on a ledge, with the vast depths beckoning beyond our vision.

'OK, lights coming on.' The giant beams at the side of the sub snapped on. Before, we had just a suggestion of the life in the darkness. This was like the floodlights coming on in a darkened Wembley. There were lots of hard and soft corals, the odd sponge here and there and a light silt covering everything, kind of like an old house that's been trapped intestate, with dust coating the grand pianos and candelabras.

It was an alien world. Everything I saw was distorted and strange. Invertebrate life abounded here: yellow rope sponges, barrel sponges you could sit inside, gorgonian sea fans that could be used by subaquatic slaves to cool their marine masters. And within these habitats were curious scrabbling invertebrates; tiny arrow crab or harlequin shrimp, bug-eyed gobies nestled with bristling soldierfish, and beyond them a big pair of curious eyes staring out of the shadows.

'This was the shoreline for the Cayman Islands during the last ice age,' Gary said. 'All the rocks you see out there are limestone, and all these little particles that you see out here represent dead organic matter.'

The sea snow tumbled down in a slow flurry, glinting in the glow of our giant beams. 'Just going to wait for the silt to settle and see what life we can find.'

Gary pointed at a strange swaying bushy structure in front of us. 'What we're looking at down there is a hairy anthem. It has these long tentacles … they grab food as it goes by. It's a member of the anemone family.'

I pressed my nose to the submersible glass, just like my little boy did when he saw his first lion at the zoo. 'Is it feeding right now?' I asked.

'Sure, those tentacles are reaching out and plucking bits of plankton … it's taking advantage of the current. It doesn't usually run this strong. They have to have the ability to survive without sunlight. Down here they have to work hard to catch the phytodetritus and sea snow that's falling down.'

It struck me that phytodetritus is a very cool way of saying 'dead plants'. I resolved to use it in a sentence at a dinner party sometime.

He pointed to another anemone-like creature below us. 'We've seen so many that we've taken to different experts, and they say, "You've found a new species." Speaking of which, I was the first person to see that species of sea cucumber there.'

Gary pointed to what looked like an aged grey turd the breadth of my thigh. I wondered if the scientific name reflected its scatological appearance but didn't ask, not wishing to disrespect his discovery.

After a painfully short time with the sub at depth, Gary adjusted a few levers, and the sub began to rise again. At about 200 feet, 70 or 80 metres, we began to get a bit of natural background blue light. As the light increased, the range of colours was just extraordinary. This was 'the sponge belt'. Giant elephant ear sponges hung in pink, orange, red and green barn doors on the wall. There were also barrel sponges a fat heifer could comfortably have had a kip inside of. Sea whips and rope sponges seemed about to lasso us and drag us into a wall that boiled with life.

We could do nothing to halt our ascent, and within the flutter of a jellyfish's skirt we were sailing above the reef and popping to the surface like a bobbing cork. When we were back on dry land, Gary handed me a polystyrene cup he had attached to the outside of the hull before we descended. The pressure had compressed all the air bubbles in the polystyrene, and it was now a perfect miniature cup the size of a thimble, like a teacup from a doll's house.

* * *

In the void we are transient visitors – humans cannot linger here. I had therefore savoured every second, seeking to imprint every element on my brain. It will always be one of the most eye-opening and formative experiences of my life. And yet we only went down to 1,000 feet – 300 metres. Three hundred! Less than 3 per cent of the depth of the deepest seas. It had felt like walking on the surface of the moon. It's a world so distant and alien from our own that it seems utterly disconnected from our reality; surely we human beings could not have had any lasting impact in this remote place?

But of course we can and do. Deep-sea fishing really began at the end of the Second World War. However, our insatiable sweep of the seas means that in the last two decades we have now started catching fish down to 2,000 metres, at depths humans have rarely visited and absolutely do not understand.

In nutrient-rich shallow and coastal seas, regular mass spawning is the normal way of reproducing. Fish live fast, producing millions of eggs with very high wastage and overcoming challenges and predators by sheer force of numbers. Down below a kilometre may be a habitat which is contiguous with the shallows, but it may as well be a different universe. The environment is extremely constant and doesn't change with the seasons or conditions. Predators are much more sparse, so reproduction and general life histories are totally different. Fish live longer, grow slower and mature later. Their metabolisms are slow so that they can survive through periods of zero food. They produce far fewer offspring – by many orders of magnitude. Fish in coastal seas generally have little or no parental investment in their offspring, whereas deep-sea fish inevitably produce much bigger eggs with greater nutritional stores and a much higher chance of each individual surviving. The movement of our fishing fleets from coastal seas to the deep, while using the same methods and catch limits, is kind of like a predator that usually catches flies suddenly deciding to catch pandas.

This is exemplified by the story of what should perhaps have been an obscure deep-sea nonentity. In the late 1970s, New Zealand fisheries discovered a new catch. Living at over 1,000 metres below the surface, they occurred in good numbers and were easy to land. There was, though, a slight issue with its common name: the slimehead. As this didn't conjure up culinary delights, the slimehead got a rebrand, and the orange roughy was born. From 1979 fishing efforts intensified and the fleets hoovered up tonnes of the newly named wonder fish.

Then, practically overnight, the orange roughy just disappeared. Too late, puzzled marine biologists started to study the fish's biology, and discovered that they have a lifespan of over 200 years and don't reach maturity until they are 40. In a fishy facepalm moment, it became clear that 85 per cent of the total population had been taken in the first five years of fishing alone.

That the slimehead bounced back at all is not down to the fact that this science led to catch limits being put in place, but purely down to the orange roughy getting so rare so quickly that it was no longer economic to target them.

The habitats too of the deep are at risk, and more precious than we ever thought possible. Although a great deal of the deep seabed may appear to be just murky sediment, it is one of the most significant repositories for carbon, and churning it up releases that stored carbon back into the water column. More importantly, there are also vast reefs of cold water corals on the seabed, astonishingly robust in their ability to deal with the cold, the dark and crushing pressure, yet utterly incapable of surviving demersal dredging nets crashing through them. One study showed that a single trawl reduced 95 per cent of a reef back to the bare bedrock, 43 per cent of the species affected were unknown to science, and many were probably unique. It has been suggested that this is the equivalent of bulldozing a village in order to harvest a can of soup from the local supermarket.

Only 5 per cent of the UK's marine protected areas ban bottom trawling, but that is largely irrelevant, as all but one experienced bottom trawling between 2015 and 2018, with trawlers spending 90,000 hours within our protected areas during that time. Forty-three per cent of that fishing was from our own UK-based fishing fleets, with 57 per cent coming from EU-based vessels.

The level of bycatch produced by this kind of fishing is extreme, but it also completely obliterates every kind of plant, algae, coral and organic substrate binding the seabed together. If you dragged a JCB bucket through your front garden, ripping up all the topsoil, you wouldn't still expect to have daisies in your lawn and hedgehogs wandering around ...

Abyss

My one single plunge into the abyss introduced me to a phantasmagoria of delights without equal. But like so many biomes, it is being plundered before we really understand it. We humans are running roughshod over the deep while it is still an undiscovered realm.

CHAPTER THREE

Seal Seas

Men Wanted: For hazardous journey. Small wages, bitter cold,
long months of complete darkness, constant danger, safe
return doubtful. Honour and recognition in case of success.

Ernest Shackleton

*Our inflatable boat bumps over greasy seas, sub-zero temperatures
starting to form a film of ice on the surface. There ahead of us is
our quarry, lying alongside a bottle-green iceberg, croc-like in its
positioning, with just nostrils and eyes above the water. I have rarely
seen such an intimidating-looking creature; everything about it
is reptilian, serpentine, menacing. While many seals may appear
as circus clowns, the leopard seal is one that will literally shake a
penguin out of its skin and has even been responsible for human
deaths. A diver who was killed here was later found with their dive
computer intact. It showed that the leopard seal had taken them
down to 80 metres below the surface.*

The animal in front of us is three metres in length and could have topped the scales at 300 or even 400 kilograms, the weight of a prize heifer. Her first approaches seem curious, as if she is investigating us, doing fly-bys and then disappearing around the back of the iceberg for a few minutes. But the longer we stay nearby, the more excited she seems to get. Her approaches get closer, her movements twitchier. She flips to show us her pale belly, then her back, dappled with the spots that give the animals their leopard name. She flashes her teeth into the camera, three, four times, yawning her mouth wide to show off her prodigious dentition, before starting to blow bubbles at us, another very dramatic threat display. She could grab my fins or feet like a cat playing with a chew toy and haul me down into the abyss; there would be nothing I could do about it.

Then a shout from the boat: 'She's just bitten the Zodiac, things are escalating, get out, I repeat, get out of the water.'

* * *

There are parts of our planet where it is routinely so cold that water is only found as ice. These places compose the cryosphere, for the Greek *kryos*, which means cold. Some of this is terrestrial: the great ice caps of Greenland and Antarctica, and the frozen tundra of Siberia, Alaska and Canada. However, frozen parts of our oceans comprise – in winter – the biggest areas of our cryosphere. They are hugely important in regulating the earth's climate, especially because of their albedo effect. This describes the tendency of white, shiny pack ice to reflect the sun's energy back up into the thin atmosphere and out into space. This is in direct contrast to the tendency of dark liquid seas to absorb that energy and heat.

One of the best-known effects of anthropogenic climate change is the retreat of Arctic sea ice, particularly in summer.

The role of the cryosphere as a thermostat for our planet is hence being brought into sharp focus, with runaway climatic feedback loops. This essentially means a change can cascade and get exponentially worse and worse. A warmer Arctic means less sea ice. Less sea ice means less reflection of heat into space. More retained heat means warmer seas, means less sea ice and so on.

It is in the cryosphere where climate change ceases to be a bunch of numbers on a balance sheet and starts to become something you can see and touch and feel. I've had the privilege of living alongside peoples in the Arctic who are custodians of an oral history going back thousands of years. They mark the passage of time throughout the year by the formation and break-up of the sea ice, and they are seeing its retreat in real time. These changes affect polar wildlife more than any other animal populations, and that particularly goes for animals who rely on the Poles for critical parts of their life cycle, like breeding and giving birth. Perhaps no animal group more exemplifies this than seals.

. . .

The 33 species of seals, fur seals and sea lions are more appropriately known by their scientific moniker of pinniped, which means 'fin-footed'. In addition, at least 50 fossil species have been revealed, although the precise path of their evolution is far from settled, with there being two possible pathways. The walrus seems to have evolved from a lineage that includes *Odobenocetops*. This bizarre animal was a bit like a narwhal, and had weird tusks that spilled out of its mouth but were held backwards alongside the body. Seals and sea lions, on the other hand, probably evolved from an ancestor they share with today's weasels and otters (mustelids). The recent discovery of *Puijila*, a superbly intact fossil of a four-legged seal with a long tail that swam in the Arctic Ocean 20–24 million years ago, has given new weight to this idea. Before *Puijila*, seal ancestors were terrestrial, first venturing into the water to find food, much the same as modern mink, fishers or martens might. In the millions of years since that change, seals have evolved to live a predominantly marine existence, with most species only coming ashore to rest and breed.

The modern pinnipeds seals and sea lions/fur seals might look superficially similar, but the latter have external ears and drive themselves through the water by 'clapping' their paddle-like forelimbs, while seals wriggle through the water using their tails for thrust. Although the Caribbean monk seal (which was declared extinct in 2008) lived in the tropics, and certain species like California sea lions and Guadalupe fur seals occur in equatorial environments, the pinnipeds are really adapted to cold waters. Their rich blubber layer provides perfect insulation, particularly for the great lumbering hulks the walrus and elephant seal. There are more species of seal in Arctic and Antarctic waters than anywhere else, and it was to the latter that I undertook a quest, in search of the most feared and ferocious of all seal species: the leopard seal.

• • •

The biggest attraction of Antarctica is surely how extreme it is – the emptiest, most remote, coldest, driest, windiest, least inhabited and highest continent on the planet. The Southern Ocean (also sometimes known as the Antarctic Ocean) is generally thought to consist of the waters of the Southern Circulation, a powerful ocean force that affects the seas south of the 60th parallel (60° south). It should be noted though that some oceanographers instead see the boundaries of the Southern Ocean as being seasonal, following the Antarctic convergence, where freezing waters from the cryosphere mix with the warmer waters of the Atlantic, Pacific and Indian oceans at their southern extent. It is also the youngest of our oceans, formed when the Antarctic continent separated from South America around 30 million years ago.

This is an ocean that is dominated by ice, with the area of coverage varying 700 per cent between mid-summer and mid-winter, but also by its main current – the Antarctic circumpolar current, which flows ever eastwards, carrying 100 times the amount of water of all the world's rivers combined. The prevailing winds follow this trend, with the roaring forties, furious fifties and screaming sixties (all named after the latitudes they occur at) the most regular and rampaging winds on earth. The high polar plateau continuously cools the air above it, creating cold, dense air that wants to sink, racing down in inversion winds that rattle off the ice cap and out to sea. These winds can be channeled by the mountainous landscape into the notorious katabatic winds, which regularly produce freezing 100mph squalls. The highest wind speed ever recorded, 200mph, was right here, and the lowest ground temperature of –89°C was measured at the Soviet Vostok Station in 1983.

The hunt for whales and seals drove the early exploration of Antarctica, and it is a familiar story of short-sighted excess. Seals had been hunted in northwest Europe and the Baltic for at least 10,000 years, while Native Americans and First Nation peoples in Canada had been hunting them for 4,000 years without any tangible effect on their populations. However, in 1776 large-scale commercial hunting arrived in the South Atlantic. The seals were primarily targeted for their dense fur, although oils were also garnered from their blubber and used for soap and cooking oil, and their tanned hides used for leather. In 1778, English sealers brought back 40,000 seal skins, and 2,800 tonnes of elephant seal oil. This was small scale compared to the industries of Newfoundland, where more than half a million seals were being taken a year, but even so, by 1830 the Antarctic fur seal (which numbered many millions before commercial hunting) was all but wiped out.

Whales were the same. The whalers of the eighteenth century targeted sperm whales in quite incredible numbers for their meat, their blubber, their baleen, which was known as whalebone, and above all else, their oils.

Once Antarctica had become a target for the trade in marine creatures, explorers wanted to push on ever further towards the interior of the continent. The South Pole is expedition folklore, with names like Scott, Amundsen and Shackleton emblazoned on the brain of every historian, explorer or romantic who's ever thumbed a well-worn adventure novel. Shackleton's extraordinary voyage of courage and fortitude has been brought into the sharpest of focus in recent years, as his ship the *Endurance* was discovered in early 2022 lying in museum-quality condition at 3,008 metres in depth on the bottom of the Weddell Sea.

Iconic as these frontiersmen were, the early exploration of the Southern Ocean and the much more significant achievement of actually finding the whole continent of Antarctica is uncertain, poorly recorded and surprisingly recent. The first sighting of land beneath the 60th parallel was by a British crew in 1819, with a further three possible sightings of what might have been the continent in 1820. It was December 1839, however, before the US Navy's five-vessel 'Wilkes Expedition' set off from Sydney, Australia, and first reported landing in Antarctica. Their landing site is still known as Wilkes Land today. Just a year later explorer James Clark Ross first passed through what is now the Ross Sea, naming the Antarctic mountains Erebus and Terror after his ships.

I've led a dozen expeditions into the Arctic Ocean, but to date have only had one trip to the Southern Ocean at the bottom of the world. Prior to that, it was both an ocean and an entire continent I had never visited, after 30 years and more than 116 countries as a professional explorer.

That all changed in 2014. Our expedition would be at sea for over a month, in the most notorious waters on the planet, making it the biggest sea journey that any of the crew had done before. All of us were nervous before departure, and despite the fact that I'd never been seasick before, I'd gone to my GP to get the strongest seasickness medication available. Among our number we had taken every conceivable anti-emetic known to science: bands, pills, ointments, ginger and tiger balm. Which in the end felt like using a Band-Aid to treat an arterial bleed.

Even to get to harbour in the Falklands was an adventure, as there were no commercial flights. Instead, we headed through several levels of military security at RAF Brize Norton and boarded a stripped-back Boeing airliner with 1980s decor, wafer-thin seats and illuminated signs warning you when to stop smoking. As we neared Port Stanley, two RAF Eurofighters fell in at the airliner's wing tips and guided us in to our landing, presumably to protect us from the watchful eyes of the Argentinians lurking just a stone's throw away across the South Atlantic.

It was in harbour at Stanley that we saw our boat the *Hans Hansson* for the first time. Originally built in 1960 as a life-saving vessel to serve first the Baltic then the North Sea, she was a sturdy-looking tub. The churning pistons of the engines below looked like a working relic in a steam museum. Dion, our captain, assured us that if they had been running for 75 years, there was no reason to think they'd break now. I was more of the opinion that I'd quite like shiny modern tech driving our props, rather than this clanking antique. The boat also seemed kind of small to be stranding ourselves in, afloat alone on the world's wildest seas.

The size of the *Hansson* turned out to be nothing like as important as its curved hull, which might have made it robust but also made it as stable as a fairground waltzer. The first night at sea was a genuine horror show. The boat pitched and rolled so violently that anything that wasn't tied down was thrown around the cabin, including everybody on board. Drawers and doors that weren't bolted shut slammed like cannon-fire. I crawled across the floor in order to put boards into the side of my bunk, and stuff either side with bags, cushions, clothing, anything I could find to cocoon me into my bed and prevent me from being thrown out with every wave. Nobody on the team slept a wink, and just after sunrise I decided to get up and get a cup of tea. After half an hour being thrown around the galley, black and blue with bruises, I gave up and went back to bed. The possibility of sleep was a joke. The idea of a month of this was a nightmare.

But then the very next day the storm subsided, the swell eased, the sun came out and the west winds fell neatly behind us. The boat was no longer all over the place like a rubber duck in a hot tub. Over a late breakfast that few of the crew could stomach, the captain, Dion, scoffed at our wide eyes. 'That was a better than average night,' he said, laughing. 'You all want to be here when it's really raging!'

There is no sense of vulnerability quite like that of the open ocean voyager. There wasn't another boat for a hundred miles in any direction. If we had succumbed to the waves and managed to get into our immersion suits, they would have kept us alive for an hour or so. But it would have taken days for rescuers to reach us – a thought best not dwelled on.

The advantage of this remoteness, though, was that the *Hans Hansson* became the focal point for everything that flew. Ships at sea often dump their rubbish, empty their toilets and spill their bycatch overboard, and much of that is free food for the birds. Boats also stir up the upper surface of the ocean as they power along, which brings fish, squid and other nutrients to the surface. In a featureless ocean, even our little tug drove the winds up and over its innocuous form, creating vortexes that brought lift to the wings of the most aerial of birds. Tiny storm petrels fluttered like bats behind us, pitter-pattering on the sea surface to draw small creatures up to the surface. Storm petrels live their entire lives way out at sea, only coming to land to breed. I've lain at night alongside their burrows on the islands of St Kilda at the far reaches of the Outer Hebrides of Scotland, hearing their bizarre call, like the odd chuckling of a jack-in-the-box. The rest of their lives is spent out here in the endless blue. I've always been moved by these birds, no bigger than the size of a starling. It's the humbling sense of fragility in the face of impossible odds; a tiny windblown bird, dancing like a ballerina on stormy seas. They are resilience against raging hurricanes, never seeking the solace of safe harbour. This is also, however, a measure of how we shouldn't perceive wildlife through a human lens. At their size, the buffeting probably provides a rewarding massage, and means they can stay aloft with little energy, a paper dart in a perpetual wind tunnel.

A close cousin also took wing at the back of our boat. If the storm petrel is the smallest seabird, this cousin is the biggest. With a maximum recorded wingspan of 4.2 metres, the wandering albatross is the largest flying bird alive today. Each wing is as long as I am tall. Compared to the wings of other soaring birds like vultures, an albatross wing looks impossibly thin and weedy. When you get up close, though, you can see that towards the shoulder there is formidable muscle, and a notched joint that allows them to lock the wing in place while soaring, to take any effort out of their gliding action. In fact the only flying creatures we know of to have ever come close to this wingspan were the pelagornithids, ancient birds that lived from 55 to 3 million years ago, and the ancient flying reptiles the pterosaurs. Our understanding of those other winged giants is very much informed by how albatrosses live today. Albatrosses can only thrive in the windiest parts of the world, as they simply cannot get airborne without driving winds underneath their wings. Once aloft, they never fly very high, but instead cruise a whisper above the waves, with one wing tip coursing the surface of the water like a sword. They can travel for many days using a system known as dynamic soaring. As winds whip the water, they are constantly redirected upwards by the waves, in updrafts. Albatrosses soar up a few metres in height using these updrafts for lift, then drop dramatically down to get speed, before rocking back up again with the wind. It's hypnotic to watch. You can eye a single albatross for hours on end without ever seeing them beat those vast wings. Students of aeronautics have observed albatrosses cruising into the wind at up to 50mph by tacking like sailboats.

Theirs is (to human sensibilities) the most lonesome life of any beast. After they fledge, they leave their nest and their parents and then don't come back to land. They just scour the deep blackness of the open sea in the search for food. They will not find a mate and breed successfully until they are around 11 years old, when they come back to the craggy slopes where they were born.

One of the very few islands where albatrosses nest was our first destination in the deep south. South Georgia looks as if a chunk of the Andes has been dropped here, with snow-capped peaks soaring straight out of the brutal waters. Southern South America is their nearest mainland neighbour, around 1,000 miles away. There is no permanent human population, and no settlements other than scientific research or old whaling stations, but it is one of the greatest places for wildlife on the planet.

We fell ashore on a desolate island after five days of battering at the far northwestern apex of South Georgia. This is Bird Island. Not imaginatively but very aptly named; at the right time of year there is a greater concentration of birds and seals here than anywhere on earth. This is the first place the west winds hit after they have whipped around Antarctica, meaning brutal winds clatter into the cliffs. Birds like the albatross rely on them.

When Captain James Cook's ship the *Resolution* first made land at Bird Island, he reported that, 'The inner parts of the country was not less savage and horrible ... not a tree or shrub was to be seen, not even big enough to make a toothpick.' As he rounded the headlands at the southernmost point, Cook for the first time realised how far he was from his goal of reaching the Great Southern continent, so named it Cape Disappointment. Now it is home to a remote outpost of the British Antarctic Survey, staffed by young (mostly British) scientists studying for PhDs and monitoring the lives of the phenomenal glut of wildlife that call this place home.

It was strange pulling up at their dock. The 20 or so occupants spilled out of the single-storey plastic buildings to meet us but then stopped short of the boat as if uncertain what to do or say. They put me in mind of a group of shipwreck victims who've been cast adrift so long they've almost forgotten the outside world even exists. They were – not surprisingly – a tight-knit group, almost like a commune, speaking their own patois, and one charismatic leader away from forming a windswept cult. Every morning they'd be up in wellies and waterproofs yomping miles over the rugged landscapes, walking transects* and counting

---

* A transect is a line across a piece of habitat used in biology to sample or measure a variable.

nest sites. Others would squat in the seal and bird slime on the overcrowded beaches, logging parasite loads and monitoring vagrant visitors like leopard or elephant seals. It took some of them three days to build up to talking to us. Some never did. A couple of the bolder girls took the opportunity on the second evening to come aboard the *Hans Hansson*, have a 'proper shower' and a beer or two (prohibited on the base), returning red-faced, scrubbed and giggling like ten-year-olds who've been sneakily scrumping in the local apple orchard.

Surprisingly few Bird Island inmates find it too much and have to leave. It's actually more common for the wilderness to infect the young field assistants with a fierce pantheism, an intensity of experience that 'normal' society cannot compete with. Many volunteer to overwinter. Others return year after year. All find it difficult to return to a modern world that seems oddly banal, safe and sanitised after a year spent in such majesty.

It was in the company of two wiry, straggle-haired and wild-eyed young biologists that we headed up onto the grassy slopes of Bird Island to find the nests of the wandering albatross. The first time you approach one of the birds, you feel a little like Gulliver in the land of the Brobdingnagians. When they stand upright, they pretty much look you in the eye, and their great hooked beaks seem as though they could shear off your thumbs. The plumage – on the adult birds mostly white, with black and dappling running down the wings – has an ethereal sheen. And what wings! When they stand before you and open them in cruciform pose, it beggars belief.

It usually takes around four years after fledging before albatrosses return to their natal slopes. We saw some of these younger birds around us, mostly trying to butt in on existing partnerships like awkward gooseberries, before being rebuffed and sent packing. It may take another six or seven years before they meet their mates for life. This strategy – almost unheard of in the mammal world – is quite common among certain birds, such as swans, geese, ospreys, barn owls and some eagles. These birds put so much effort into provisioning for their few, large chicks, that they absolutely need the efforts of two committed parents. In addition, they may spend much of the year on long-distance migrations so have no time to waste on the lengthy process of finding a new mate. They get started early in the season and see things out together as a couple. Albatrosses can live to be more than 70 years of age, so this is a relationship that can last decades.*

Once coupled they return every two years to the same grassy slopes and set up their nests in the same locations. Upon meeting their partners again, they engage in a duet of incomparable beauty. The larger male approaches with his head down and back humped, waddling from side to side with exaggerated swagger. The female comes to meet him, and they both spread their wings as if about to go in for the mother of all cuddles. Then the male throws his head back, sky-pointing with his giant beak, before sounding a resonant, clarinet-like call. They tenderly touch each other's bills, preen each other, and then repeat the ritual. All over the hillsides, the same duets take place, while in nests a giant brown raggedy mess of a chick sits waiting for its parents to return and regurgitate squid sick for it to glug.

It takes about 90 days for the chick to fatten up to a weight where it can attempt a flight. The giant goose-like chick will scrabble from the nest, turn downhill and into the wind, open its huge wings and set off on the most ungainly sprint you have ever seen. Even experienced adults will crash and collapse more often than not, so watching the youngsters do it is pure comedy.

---

* Some female albatrosses who don't find a mate may engage in extramarital copulation, and then form bonds with other females, sharing the duties of rearing their chick together.

Eventually, though, they manage it, and once airborne they won't return for at least four years.

GPS trackers placed by researchers from Bird Island show that males and females forage in different areas of the Southern Atlantic and cover vast distances. While some birds fly circuits off the coast of Argentina, other individuals cast themselves onto the trade winds and literally fly around the world. One grey-headed albatross achieved this in 40 days, travelling 13,670 miles without landing.

Inevitably, though, there is a sad human intrusion into the story of the world's most romantic bird. Albatrosses are not great divers and land on the sea to snatch squid from close to the surface (even feeding on the remnants of squid vomited or wasted by whales). Increasingly these wanderers of the roaring forties are mistaking ocean plastic for squiddy morsels. Albatrosses that wash up dead on Southern Ocean beaches rot away to reveal twisted lumps of thousands of pieces of plastic where their stomachs once were.

Longline fishing (where a boat will tow perhaps 60 miles of lines with thousands of baited hooks) is even more lethal. They're mostly targeting high-value fish like tuna but produce an unimaginable amount of bycatch, from dolphins to turtles and sharks. One of the biggest losers, though, are albatrosses, which spot the lures being dragged beneath them, then duck down to grab the food and get hooked. It seems the slightly smaller female albatrosses are more likely to snatch the shallow hooks. It's mostly the males that return to their nests, and then stand vigil, waiting to be reunited with their partner, gazing to the west winds for a love who will never return.

An estimated 300,000 birds a year are drowned like this, and it shouldn't be happening. The process has been understood for decades, and all the longliners need to do is weight their hooks and provide bird scarers on the lines – these reduce bird mortality by up to 99 per cent.

. . .

The real metropolis of Bird Island is the thriving shore. Although Antarctic fur seals were once hunted to the edge of extinction for their dense fur coats, they can now be found here in their hundreds of thousands. They gaze at you with dark puppy-dog eyes, before roaring and lunging at your feet, baring their lion-like teeth and trying to tear a chunk out of you. One of the most impressive predatory spectacles you could ever see is these fur seals thundering out of the surf to catch hold of a gentoo penguin, and then ripping it apart in a gory explosion of white water, blood, feathers and squirted penguin poo. We emulated the researchers and carried sticks to keep the seals and their fangs at bay.

Because fur seals have a hefty layer of fatty blubber to keep them insulated as well as their dense fur, they can't stay on land too long or they overheat. Many of the animals were therefore lolling in the shallows, oblivious to the fact that the water was hovering around 0°C. For us to dive with them we needed specially designed drysuits, with tight cuffs to keep even the merest dribble of water from penetrating inside. Under these we wore all-in-one babygrows filled with polystyrene beads; the air bubbles within them would be warmed by our body heat. Still, thick neoprene hoods would not ward off the penetrating chill, and there are no gloves that can stop the cold stabbing into your fingertips. In the shallows, in green waters with limited visibility, the seals seemed to be mocking us. All our technology was no match for fat and fur.

Despite all the advantages of biology, the seals also have a hard time of it. Some years, only one in ten seal pups make it to adulthood, with many dying of starvation or being battered or crushed by rampaging males on the rocky beaches. This far south there are very few bugs to do the job of clearing up, and with no vultures, jackals, hyenas or other large scavengers, several seabirds have stepped into this ecological niche. The Antarctic skua is a big bulldog of a bird, known for mobbing anything that gets too close to its ground nest sites. They flock around a baby seal as soon as there is any sign they might be vulnerable, and will peck at their eyes and anuses, trying to speed up the inevitable.

The giant petrel is bigger still, and looks like a cross between a dodo and a waddling dinosaur, with a huge beak straddled by two tube nostrils. These give them a keen sense of smell, as well as the ability to shed the excess salt they get from their food and seawater. Giant petrels plunge their heads into the seal carcasses, emerging besmeared with blood and goo. The researchers refer to them as 'GPs', 'stinkers' or 'shit chickens'. They are distantly related to albatrosses but can be distinguished by the fact their tube nostrils join over the bill whereas albatrosses' are separated. And while the aloof albatrosses stay distant from the melee of the beaches, the GPs scrabble over the stacks of seal sausages with gusto, the malodorous marine mammal mosh pit forming the basis for an entire festering food chain.

· · ·

On leaving Bird Island our boat now began the journey south, hugging the coast of South Georgia to gain as much protection as possible from the battering of the winds. The boat sailed overnight and we awoke to otherworldly stillness as we made harbour at Grytviken. For the first time in ten days we stepped outside to see waters as still as glass.

It was the shelter from the prevailing winds that had made this place so important to the first major economic exploitation of the south. From 1904 to 1964 this station was perhaps the biggest outpost of one of the biggest whaling industries in history. Southern right, blue, sperm, fin, minke and humpback whales were caught in the waters around South Georgia then brought back here for processing. A staggering 175,000 were killed in the immediate waters, and 1.4 million in the Southern Ocean as a whole.

Before leaving for the expedition, I'd pulled some film from the BFI film archives, juddering black and white images of men clustered in sou'westers, with thick beards and blackened faces, tarred by soot from burning rendered whale blubber. They fired cannons with explosive harpoons from the bows of their ships, before winching the hulking corpses on board, or dragged them alongside and into port. Long-handled flensing blades would then be used to scythe through the blubber and expose the meat beneath. It was a horrible life for them, but one which bore rich rewards, especially for the companies or countries that sent those fleets out into harm's way.

Operations ceased in the 1960s as the industry was no longer commercially viable – in truth, there were no whales left to kill. Grytviken was deserted, and left as a kind of epitaph to those terrible times. Rusting scuttled ships, rusting vats and barrels, rusting machinery and great rusting chains hanging off great rusting anchors. It was like the Southern Ocean was thumbing its nose at the best efforts of the hardy men who battled here. 'Look on my works, ye mighty, and despair.'

Among a few jolly red-roofed white buildings were vast melting pots, where the whale oils were rendered down to make margarine and machine lubricants. The bones were ground to be used as fertiliser. One simple sign hit me right in the chest. It said that 30 whales could be processed there in a day. The biggest of these was a single female blue whale who was 34 metres long and getting on for 200 tonnes. I already knew this whale well. I'd described her measurements in dozens of lectures and programmes over the years. This female blue whale was the biggest animal that has ever been recorded on our planet, more massive than any of the known dinosaurs. She is essentially the yardstick by which we measure all large animal life, and she was killed there in 1912 to be made into margarine, bonemeal and fertiliser.

Much of your experience of a place may be defined by the weather. Many have described Grytviken as fascinating, even beautiful. Maybe they had blue sky and more open eyes. To me, visiting on a grey day, Grytviken was a seedy reminder of human greed, excess and short-sightedness. On my first solo expedition in 1990, I wandered around the Killing Fields in Cambodia, the notorious death camps of Pol Pot and the Khmer Rouge. The fields have long since been cleared up and turned into a questionable tourist attraction, but back then bits of skull, skeleton and clothing lay around in the shoddily excavated shallow graves, a grisly testament to a terrible period in history. You could feel a sombre and unsettling sense of suffering there. I felt that same sense at Grytviken too. Everywhere you looked, seals and king penguins perched on the great skulls of right whales, and in among the ribcages of ocean giants. It felt like a place where great suffering had happened, not all of which could be cleansed away by wind, rain and time.

Nowadays, Grytviken is a place of pilgrimage for lovers of the history of exploration, as it is home to the grave of Sir Ernest Shackleton. I'm sure the small graveyard here is home to many other heroes who lived lives of adventure and whose stories are tinged with a heavy dose of heartache. One grave, though, stands out, with a thin upright granite block, and Shackleton's simple epitaph. Seeing this understated monument, here in this godforsaken place, pointing south towards the Pole he so coveted, was a surprisingly emotional moment for me. Shackleton is perhaps most famous for his failure to cross the great southern continent and becoming trapped in the ice with his crew for years on end. However, I believe everyone should at some time be required to read *South* (Shackleton's own account of the expedition) or Alfred Lansing's *Endurance*. Put to the sword in the most intense conditions our planet has to offer, the crew took their suffering in their stride, exhibiting a selfless fortitude that is truly humbling. Many times on expeditions in the past, when I've felt low or beaten, I've thought of Shackleton and his men to give myself perspective. They were the living embodiment of a great truth: that selflessness, looking after each other, positive thinking and just getting on with it is the true key to survival. And after they had endured hardship that is difficult to fathom, the crew returned to Europe, and most were shipped straight off to the trenches of the First World War, without glory, or – it seems – complaint.

• • •

In popular perception, the Arctic and Antarctic are barren places, frozen 'wastelands', where everything battles to survive and living creatures are few and far between. We as humans evolved in the warmth of sub-Saharan Africa, and even with the developments of today's technology, the conditions in the cryosphere seem 'extreme'. However, in the context of our planet and its living history, the Arctic and Antarctic circles are not actually extreme environments at all. Tens of thousands of species of animals, birds and marine creatures have evolved over millennia to be comfortable in cold climes. For any such organism, the conditions of the polar regions are not just something they endure, but something they require. Thriving marine food chains, driven by constant upwelling from the depths, lead to vast amounts of primary production in the form of soupy plankton. The plankton feeds Antarctic krill, which swarm in shoals that might number 30,000 individuals per square metre. In total, there could be as much as half a billion metric tonnes of krill in the ocean, equating to hundreds of trillions of animals, making them perhaps the most numerous potential food source on earth.* Whales, penguins and fish feed on krill, as do seals, with two species being especially adapted to take advantage of this bounty. The first is the leopard seal. Though their vicious canines are purpose-built for hammering warm-blooded prey, their molars are tricuspid, meaning they include three cusps that function like a sieve. The leopard seal can swallow mouthfuls of seawater and krill, then flush the water out and keep the krill within. Crabeater seals go one further with pentacuspid (five cusped) teeth, and at times feed on nothing other than krill.

This abundant food chain means that our cryosphere is home to more great gatherings of living organisms than anywhere else. One of the most spectacular of these is St Andrew's Bay, our last stop before departing South Georgia. This broad, sweeping black sand beach has a river running through its centre, snow-covered peaks looming over it, and the noisiest, stinkiest and most beautiful clientele you could ever imagine. This is the biggest colony of king penguins in the world. Officially there are 150,000 breeding pairs, though that count was made 30 years ago, and most observers believe numbers have doubled since then to well over half a million birds.

It was totally overwhelming. On every square foot of dark sand, kings stood like glorious gaudy bowling pins. When we approached, they would waddle up to us, extending their necks

---

* Even krill are now being targeted by our fisheries, caught for fish and animal food, to be ground into fish oil, and to be used as fertilisers.

to give us a good look. So far on our expedition we'd already spent the night in a macaroni penguin colony, and lain in among a rockhopper penguin colony to watch them bounce down the slopes alongside us. Both were chaotic and stinking, we were constantly being nipped, and the discordant noise set our teeth on edge. Being among a king penguin colony, though, was like sitting among the woodwind section of a 100,000-strong orchestra, constantly in voice with beautiful bugling calls.

Each individual king penguin is a wonder of evolution. Many millions of years ago, the penguin's ancestor would have been a flying seabird. Its wings would have been a considerable hindrance for diving, however. (The wandering albatross, for example, with its vast wingspan, cannot dive at all, and can only duck its head underwater.) Gradually, over millennia, the bird's wings would have shortened and stiffened. The joints inside the wings fused at the elbow and wrist, and the bones flattened and became stout and hefty. The wing became a paddle, more like a seal's front flippers. Estimates differ as to when this might have happened, but it seems there could have been giant penguins as tall as humans that lived alongside the last of the dinosaurs. In losing the power of flight, evolution could be said to have gone backwards, but in actual fact this adaptation allowed penguins to become the greatest swimmers and divers of all the 10,000 species of bird.

Kings rival emperors for the title of the most beautiful of all penguins. They have ivory white bellies and silver grey backs, the simple countershading camouflage shared by most marine predators – seen from below their bellies blend into the white sky, and seen from above their dark backs disappear into the blue of the deep – but their heads, which stick up out of the water as they swim at the surface, are flamboyantly coloured, with ink-black faces and a spoon of orange about the ear patch, and their upper chests are golden yellow. We'd seen a few individuals over the previous weeks and had sat and marvelled at their beauty. To see half a million in one place was more than my senses could take.

The waters around St Andrew's Bay are inevitably cloudy with glacial silt and the squirty guano of half a million penguins, so we had written off the possibility of diving in them. However, on our second morning we were woken by a call from the crew: 'Guys, guys, come and look at this!' We clambered out of our bunks and ran out to find the *Hans Hansson* surrounded by thousands of bobbing black and gold penguins, ducking under the water to allow droplets of icy quicksilver to course over their slick plumage. The still morning air resounded to their bugling calls, and the rafts of penguins at the surface were so close to the boat you could have picked one up and walked off with it.

'What d'you think, Johnny?' I asked my underwater cameraman.

'It's unreal,' he replied in his thick Belfast brogue. 'Are we allowed to get in?'

We looked toward captain Dion, who shrugged. 'Well, no one's dived with them before, but that's more been cos there's been no viz …'

We peered over the side. The water looked like it had been laced with an excessive dollop of lavender bath salts … dense blue.

'Well it ain't crystal clear,' said Johnny, 'but it's penguins, right?!'

The penguins were heading out to feed and would not be around long. Unprepared for diving, the team thundered around the boat, desperately trying to get into the water before it was too late. After we hit the icy water, it became clear that there was probably only a metre of visibility, but that didn't matter, as the penguins were literally coming up and pecking on the camera lenses, perhaps seeing themselves in the glass.

Adapted to hunting underwater, penguins are clumsy on land but transformed beneath the waves. Their torpedo-like form is stunningly hydrodynamic, and their sleek, thin feathers both insulate them in the freezing water (they also have a layer of heavy blubber that would be too heavy for flying birds) and aid them in their streamlining.

Having not long left the beaches of St Andrew's Bay, some air was trapped within their feathers, and as they powered through the water, streams of tiny bubbles followed them like vapour trails behind a fighter jet. They looked like dumpy tracer bullets, fired out into a twilit sky.

The beach at St Andrew's Bay is also a haul-out for the world's largest pinniped, the southern elephant seal. It's impossible to believe that these corpulent slugs are actually voracious predators, as they lie groaning and heavy breathing on the shore. Their rheumy eyes are slick with spittle, snot dribbles from their nostrils, and great sheets of skin peel off their bodies as they lie sweltering in their own faeces. They are about as appealing as Jabba the Hutt. However, like the penguins, they are transformed once in the water. In fact this is probably the closest thing alive today to the proto-whale, which would have come to land to rest and breed, but spent the rest of its days submerged, its vast bulk buoyed up by the water.

These morbidly obese maggots lie slovenly in the sands, but the need to breed turns them into titanic tyros. Elephant seal males battle to become beachmasters, commanding mating rights over a harem of females in one of the great gladiatorial spectacles of the natural world. They use great tusk-like canine teeth as weapons and slam their heads into the blubber-laden armour of their foes. To enhance their success, the male elephant seal has evolved one of the greatest examples of sexual dimorphism. The female is hefty at 200 kilograms; the male is a monster at 6,000 kilograms, or six metric tonnes.

While we were wandering on the beach among the penguins, one of these battles broke out between two males, scattering females and squawking penguins everywhere as they thundered towards each other. Suddenly there was a gargling roar that sounded like an old Royal Enfield motorbike with a hole in the exhaust revving up. Steam poured from their mouths and nostrils as the combatants wormed towards one another at speed, then raised up to height. Two sumo wrestlers going head to head, they then hurled themselves into one another. Blood and spit flecked their mouths, and their chests, criss-crossed with scars, turned pink then red with blood. It was astoundingly violent, the thuds and roars carrying right across the bay. Eventually, one male succumbed and was vanquished, heading out into the waters to cool down and lick his many wounds, while the new beachmaster roared his victory, like Rocky at the top of the steps in Philadelphia.*

· · ·

In Shackleton's great escape, he and his crew battled eastwards from the Weddell Sea to South Georgia, and finally made his break for freedom across these mountains. We would be going in the opposite direction, heading south and west towards the Antarctic peninsula, and setting out to sea for another six days. Where Shackleton had just small wooden row boats, we would be safe and warm within the *Hans Hansson*, and we were now of

---

* When filming northern elephant seals in Año Nuevo, California, we saw one huge beachmaster that had a circle of plastic packing tape noosed around his neck. It was so strong that the seal had grown without being able to break it. The plastic had cut down a hand's span into the blubber of the animal.

course old hands at big seas. Our time on land and in sheltered harbours had given us respite. Even Kiri, our most seasick crew member, had managed to eat and regather some strength. The way ahead held no fears for us. And then captain Dion came into the galley with his face grey and strained. We were used to him scoffing and pouring scorn on our reactions to the Southern Ocean swell. But not this time.

'Guys,' he said. 'I'm not going to sugar-coat this ... prepare for hell.'

Dion ran through the charts and the weather predictions for those who could bear it. Kiri headed straight for her cabin to start readying herself for the worst. From here it was perhaps 800 miles to the Antarctic peninsula. Winds were expected to be gale force, gusting to near 100mph. We would need constant night-time vigilance for icebergs, but the decks would be out of bounds to anyone not wearing an exposure suit.

No matter how much we had steeled ourselves for what was to come, it was to be one of the worst experiences any of the crew had ever had. The wind howled around our tiny tub as if trying to dismantle us; the noise alone could drive you to madness. The brutality of the battering as the waves hammered our hull would throw you around inside with such violence that you'd sometimes be completely weightless, like an astronaut hanging in the air as the floor disappeared from beneath you. Every case or piece of furniture that wasn't tied down would turn into a bucking steer inside the cabins. To get to the head (toilet) was pretty much impossible, but then drinking was too so there wasn't much need to pee. Somehow, throughout this all, Dion and his crew stood at the wheel, in what was by all accounts a frightening experience even for them.

Day five was the worst, with the blackness of the night bringing a force ten gale – the same as the great storm that hit the

UK in October 1987. Back then I can remember waking to light streaming in through my curtains; we lived in among ancient oaks, and every single tree that surrounded my house had blown down. Out here there were no trees, but the maelstrom blew mighty waves into towering whitecaps higher than the roof of my childhood home. They crashed clear over the top of our boat. I have never felt so small, so utterly exposed, so totally helpless.

Crew members had their exposure suits ready to go. Sat phones were charged. In our heads we recited emergency channels and rehearsed SOS protocols and our route to the lifeboats. Every one of us whispered a prayer or two. There are no atheists in the Southern Ocean in a force ten gale.

How the *Hans Hansson* made it through the tumult I don't know. But somehow the tub survived the storm, and the morning of day six rewarded us with wind-washed powder-blue skies and a gentle swell that allowed us to creep from our bunks, find something to eat, and finally have a poo after having been plugged up like hibernating bears. A giant iceberg like a fairy castle bobbed in front of us, with a small huddle of Adélie penguins standing on a low shelf. Around it was a blue Slush Puppie of ice, with broken bits and defrosting chunks. In among it, three humpback whales circled, blowing great bushy spumes of spray from their blowholes, before spyhopping to the surface to give us a good look. It was as if we had awoken from a particularly fierce nightmare to find ourselves in paradise.

Like most visitors to the Southern Ocean, we had made our way to the Antarctic peninsula, the curling tail of the continent, projecting nearly 1,000 miles up towards the tip of South America. The peninsula comprises half a million square kilometres of mountains (which are considered the tail end of the Andes, with a submarine ridgeline connecting them), 80 per cent of which is permanently covered with ice. It has long been the subject of claims for sovereignty from a variety of nations, who have dotted research stations along its length (including the UK's Rothera), but the Antarctic Treaty ensures no one attempts to enforce their claims.

It is also the part of the continent most affected over the last 50 years by anthropogenic climate change, which has had significant effects on sea ice formation and the success of species that rely on it, like Adélie penguins. In the early 2000s, the peninsula was estimated to be losing 60 billion tonnes of ice a year.

Sea ice was at its minimum when we were there in February, reaching its maximum in September – again almost the polar opposite of the Arctic. The ice here in the Antarctic has a broader range of influences than that to the north, and because of that the sea ice is more variable year on year. In the Arctic the trend is for sea ice to be universally dramatically shrinking, but the Antarctic picture is more complex. NASA first began tracking the sea ice extent from satellites in 1978, and from then to 2017 the maximum winter freeze appeared to increase in area. Locally some areas decreased in size, but some of the seas of the Southern Ocean showed increases or no change. Though climate science deniers may seize on this as a sign that climate change isn't real, the truth is that the present extent of sea ice is not the only factor at hand. The Antarctic ice sheet is steadily losing mass, and large-scale events like the collapse of the vast Larsen B ice shelf represent the loss of hundreds of thousands of years of ice accumulation in a matter of years.

Temperature changes are also tricky to pin down. The scientists of NOAA (the National Oceanic and Atmospheric Administration) show that while parts of eastern Antarctica are registering colder temperatures, this century the peninsula has increased in temperature by an average of half a degree a decade. But in more recent years it has started to cool again. One of the greatest challenges about climate change for any broadcaster, conservationist or storyteller is that it is so vast, so amorphous, so impossible to generalise about or provide simple evidence for cause and effect. It's also so easy for the naysayers to throw doubt on. 'But Antarctic ice isn't shrinking!' Tucker Carlson might shriek, as if he has discovered the crux of the debate all on his own, and as if the world's greatest climate scientists didn't know that already.

This complexity is the main reason we are still having arguments as to whether the climate crisis is even happening. It's especially ironic when the climate sceptics turn to the Antarctic peninsula as fuel for the fire of their doubt, as the most advanced global climate science has arguably occurred right here. As the most accessible and visited portion of the continent, it is also the best studied. Research into the stratigraphy, glaciology and palaeontology of the area, plus studies of fossilised trees, plants and their pollen, has allowed scientists to understand the changes of the last 100 million years in detail, showing that at times this was a bountiful subtropical basin, alive with flowering plants and palm trees. The study of ice cores from the ancient ice sheets help us to analyse how carbon dioxide and other gases and particulates in the atmosphere have tracked or driven falls and rises in global temperatures.

The peninsula is also by far the most wildlife-rich part of the continent, seasonally home to millions of seals, penguins and cetaceans. None of these feed on land, or get any sustenance from terrestrial ecosystems other than safety when they are at their many breeding colonies. Every single one of these animals is fed from the bounty of the Southern Ocean. Not surprisingly, then, this is the most coveted destination for any hardcore scuba diver.

In winter months, you need to drive a big Archimedes screw down through the many metres of pack ice, in order to plunge into waters so clear you appear to be flying. But we were here in the summer, as much of the ice fractures and falls apart, setting in motion a chain of events that leads to significant algal blooms in the water. Much of the time, the sea is the colour of my mum's leek and potato soup, and you can barely see your hand in front of your face. The temperature is generally around freezing, although it can be as low as –1.8°C before it actually starts to freeze because of the salt content of the water. This plays havoc with our equipment – especially our regulators, which have to be encased in condoms of oil to prevent them icing up (which would cut off your air supply while you're submerged). As we didn't have the space to store all our dive gear inside the *Hans Hansson*, it stayed rigged on the scuba cylinders outside on the RIB through the worst weather of the polar night. Nothing makes you feel less like diving than stepping outside to see icicles forming all over your equipment.

To counteract the cold, we had custom-made drysuits, with thick thermal underlayers beneath. On dive days, we'd be in our drysuits from morning till night, so the set-up included an open-ended condom that fitted over our bits, leading to a tube and a valve which allowed you to pee underwater. I was a novice to this system and tried it out on my first sub-zero dive. To begin with it seemed everything was going swimmingly, then I felt a build-up of pressure where I'd obviously kinked the hose. The pressure backed up, then there was an inevitable release, and a pleasant warm feeling flooded into my lower drysuit. I was mid piece to camera at the time, so I interrupted my flow to say to the cameraman: 'Johnny ... I've just tried out my pee valve ... with disastrous consequences.' He rolled his eyes. All was fine for about five minutes, till the warm feeling subsided, making way for utter freezing cold. We aborted the dive, for me to pour yellow slurry out of my drysuit.

The second dive was even worse. The location was an amphi-theatre of towering mountains called Paradise Bay. All around the cove, powder-blue and white glaciers slid slowly into the sea, forming petticoats around the flanks of the peaks. Icebergs of every hue of green and blue glowered at us, shining aquamarine, dead white that seemed to suck up the light, cobalt that you could lose yourself in. It is a painstaking job kitting up for one of these dives, with brutal winds chilling us to the core before we could even step over the side. However, no sooner were the team in the water than the dive supervisor came over the dive comms: 'I don't like the look of that berg.' There was a brief pause, before his comments became orders.

'Divers, abort, repeat abort dive, get to the surface as soon as possible.' The team were dragged out of the water and into the RIB, and we gunned the engine as an iceberg the size of a family house ground past us, and right over the top of where we had just been.

Our main target in the peninsula was a dive with the most notorious and sinister predator of these green waters. In the shal-lows close to the penguin colonies, one polar animal is queen. I heard our first one before I saw it, a forceful out breath that resounded around the icebergs.

Obligate carnivores, they menace anything hanging out around the ice edge, ambushing seals and penguins. All around us were signs that the leopard seals ruled here – for example, we saw crabeater seals with distinctive leopard seal teeth scars running down their sides. The possible presence of leopard seals alters the behaviour of every small penguin species. Gentoos will stay ashore until they are near starvation if they fear a leopard seal is patrolling their colony. When they finally do run the gauntlet, they walk into the water, putting their face down into the waves to try to see if there's one around. When they finally commit, they sprint out through the danger zone in numbers and at top speed, porpoising out of the water as they try to overwhelm and distract the hunters.

The largest leopard seal females are getting on for half a metric tonne in weight, which means only elephant seals and walruses are bigger. In 2003, a female marine biologist working here on the peninsula was the first person to be killed by a leopard seal. Accounts suggest she was swimming at the surface when the seal became agitated, grabbed her, and took her down to 60 metres in depth, drowning her.

The first leopard seal we saw cavorted alongside us, then plunged straight for our RIB, creating a bow wave like the parting of the Red Sea. It put my heart in my mouth. Leopard seals hold themselves at the surface much like crocodiles do, with their bodies hidden but the senses exposed, their long necks twisting and turning in a sinuous, almost reptilian manner. This minimal exposure strategy hides them well from the eyes of their prey. Looking at the skull and dentition you'd assume you were looking at some ancestral big cat. They have vast canines, and huge, heavy, bony skulls. The names sea lion and sea leopard are apt indeed.

The sense of the unknown for the dive team was powerful. All of us had been diving with sharks for decades and felt as though we'd learned the rules of how to behave in those encounters through experience. But none of us had ever dived with a leopard seal, and it was a plummet into the unknown.

The next day, we slowly pootled the RIB in among the small bergs, gentoos leaping around us, with Arctic terns, at the end of a journey that had seen them truly span the planet from Pole to Pole, calling above us. First a medium-sized leopard seal buzzed our boat, making as if to bite at our outboard motor. Then we saw another animal propelling itself through the shallows towards a penguin, narrowly missing its target.

Below the waves was a different world. We swam underneath the length of a jade green iceberg, cautiously heading into the seals' territory, keeping as close together as possible to create an imposing-seeming group. The berg itself would have originated as snow falling high on an ice cap at least 1,000 years ago, but in some cases as many as a million years before (in marked contrast to the northern hemisphere, where Greenland ice 'only' gets to be 100,000 years old). Crystals on the berg surface twinkled like stars as my torch swept across them. This is an imposing place to be; icebergs can crack, break or roll, creating extremely dynamic environments, and you wouldn't want to be anywhere near them when that happened.

My heart was pounding out of my chest. Normally I wouldn't think of swimming into an animal's world if I was this nervous, sending trembling signs of stress out into the water with every heartbeat. But too late, there she was, lying on a submerged ice shelf just metres in front of us. At our approach, she slipped off the shelf and was gone. For a moment, it seemed for good. But then she returned, swimming past us, turning up to look at us with her big dark eyes, grinning with her sinister smirk.

For the first few minutes, she just repeated the swim pasts, turning on to her side to eye us up and down. Then, as she gained in confidence, she'd reach up to nip a flipper, and started to twist and turn in pirouettes, showing off her skills. First she'd show her pale underbelly, then the spotty dorsal surface that gives the animal its name. Her head started to nod, then she extended it out like the neck of a huge ice snake. The longer the encounter went on, and the more we built our confidence around her, the more she started to build in confidence too; however, it felt less as if she was learning to be friends with us, and more as if she was getting excited, manically so.

Now when she passed, she would extend that concertinaed neck and flex her mouth open, demonstrating her fearsome teeth in an exaggerated yawn into my camera lens. Then she started to blow bubbles at us. This is – if anything – a worse sign than the yawn. It's an unmistakable dominance display I've seen used underwater by a dozen different seal species, and is usually a precursor to a charge or worse. There have been a few times diving when I've honestly felt that I've made a big mistake, and this was one of those times. There was something about her that just felt wrong. It was as though she was working herself up to do something vicious.

The underwater comms crackled into sound. 'She's just bitten the Zodiac [our inflatable boat], things are escalating, get out, I repeat, get out of the water.'

We didn't need telling twice. As a unit, we swam away from the security of the iceberg and out toward the RIB. Now, however, we were at the surface, our fins pointing down into the water, and our leopard seal invisible beneath us. We all felt very vulnerable indeed. It's no easy feat to clamber out of sub-zero waters wearing full drysuits and double cylinder systems, but somehow we managed it, every second expecting a set of jaws to clamp on a leg and drag us down into the abyss.

. . .

Under the Antarctic Treaty, the entire continent of Antarctica is dedicated to peace and science, with the 1991 Environmental Protocol adding extra protection for the cold land mass. This is supposed to be the one place on the planet we cannot denude, cannot exploit and cannot pollute. For the moment, that grand ideal stands firm, but it is surely on a knife-edge. We may champion the sanctity of the forgotten continent, but as rare minerals and metals become more valued and more scarce, that ideal might soon be challenged. Will we still believe in keeping Antarctica sacrosanct when the obscure metals that run our smartphones run out and the resources to replenish them can only be found under a faraway ice sheet? What about increased tourism? Bio-prospecting? As fish stocks run dry in the rest of the world's oceans, the temptation to head further and further south with nets and lines will surely become too strong. Although a limited amount of highly regulated fishing takes place here, illegal fisheries are already plundering the Southern Ocean, far away from the prying eyes of international observers.

But despite all the looming threats, the Antarctic remains our planet's greatest wilderness, with untold miles where humans have yet to tread, and with untold marvels beneath its frigid seas.

CHAPTER FOUR

Polar Predator

For humans, the Arctic is a harshly inhospitable place,
but the conditions there are precisely what polar bears
require to survive – and thrive. 'Harsh' to us is 'home'
for them. Take away the ice and snow, increase the
temperature by even a little, and the realm that
makes their lives possible literally melts away.

Sylvia Earle

*She sniffs the air, black nostrils sucking in scents that may have
travelled a dozen miles on the sterile Arctic breeze. Perhaps she is
smelling me – after all, we are standing mere metres away from
her. More likely though it is the slovenly pile of walrus, lying corpulent as giant brown and pink maggots on the gravel beach nearby,
that she can smell.*

*The polar bear drops her head to the ground, and picks up a
piece of seaweed. She chews on it distractedly. Polar bears regularly*

engage in so-called displacement behaviours; activities that seem out of place in a particular situation. It's usually when they're stressed, mystified, or torn between two plans of action. Or perhaps this one is just hungry.

I can tell she's female from her proportions, lacking the massive head of the males, and with a stouter hind section, giving her a swaying gait. There are also tell-tale signs that this girl has been having a tough time of things. Her belly and rear legs are dirty and bedraggled, and she bears two long dark wounds down one flank. The wounds are the same distance apart as a walrus's tusks. Perhaps in her desperation to feed, she took on an animal that was above her pay grade.

Having been stalked by very hungry bears before, I know to be incredibly wary. A desperate bear is not to be trifled with. Even underpowered and starving, this is an animal that can kill a beluga whale with a single cuff of her giant paw. I've seen a bear dragging a tonne of bearded seal out of the water and up onto an iceberg to eat it.

Again she chews on the piece of seaweed. Then she turns slightly away from us, as if to convince the crew and me that she could not be less interested in us. She then sways her head from side to side and makes for us, determined strides covering the distance in seconds. If she had broken into a run she would have had us. Instead the motor of our boat guns into action, and we push off from the shore. We breathe again. She is so close it seems I could reach out and stroke her scarred snout.

• • •

The Antarctic is a continent of land surrounded by ocean. At the other end of the planet, the Arctic is the exact opposite: a land-free basin filled with water that freezes, roughly surrounded by continental landmasses. Together they comprise the cryosphere, but the Arctic Ocean is vastly different to the Southern, firstly and most importantly because of that lack of landmass at the centre of it. Surrounded on all sides by land made up of the northernmost extents of Alaska, Canada, Greenland, Scandinavia and Russia, it is subject to wildly different weather systems to down south.

While people have been able to walk to the Pole since Cook in 1908 and/or Peary in 1909,* they do it over frozen sea – constantly shifting pack ice, which moves with tides, winds and currents, often carrying polar explorers miles and miles in the wrong direction. In the Arctic there are no penguins, though the (extinct by human hands) great auk was a close equivalent. Certain populations of the white gyrfalcon – the largest species of falcon – live and hunt only out over the pack ice, arguably making them a marine bird of prey. We've even filmed snowy owls standing out on the ice floes, though what they're hunting there is anyone's guess!

While you could argue that the orca tops the food chain in both polar environments, the marine mammal that occupies the equivalent ecological niche of Antarctica's leopard seal is also the true icon of the North: the polar bear. It might seem strange to think of a bear – classic terrestrial carnivores – as being a marine mammal, but even its scientific name, *Ursus maritimus* (the marine or maritime bear), belies this fact. Some bears never set foot on dry land in their entire lives. No animal epitomises this region and its struggles more than the great white bear.

Polar bears are relatively young in evolutionary terms, having evolved from the older brown bear around 480,000 years ago. It seems likely that they migrated north during a warmer period

---

* Unlike the South Pole, the first person to reach the North Pole is shrouded in mystery and controversy even today. Perhaps the first genuine and certain person to reach the Pole was Wally Herbert, who walked there in 1969.

of geologic time and then became isolated. In the interim, their fur changed from brown to white (though their skin remains black), they developed giant snowshoe-like feet, and a thick layer of fatty blubber to aid their insulation. Their two layers of fur are so effective at keeping them warm that they have to be incredibly careful not to break into a run too often, as even in −40°C they'll rapidly overheat. Despite their diet being laden with heavy fats from seal blubber, they have no build-up of fats in their arteries. This appears to be down to genes that aid the transport of fat in the blood, and could well be vital in the treatment of high cholesterol and heart disease in humans.

Every time I take an expedition into the Arctic to find polar bears I see a different side to their character. On the north shore of Alaska I've seen bears scavenging from the remains of bowhead whales (hunted by subsistence whalers and then dragged onto the shore). Through the compacted ice are the giant footprints of the bear, and alongside them the tiny paws of Arctic foxes that follow them for entire seasons like servants, sneaking in to pick up the scraps the great bears leave behind. In the Scandinavian north I've witnessed the formidable sight of as many as 80 bears gathering to feed on a stranded sperm whale carcass. Fights between these usually solitary and temperamental animals are rare in these moments of abundant food resources.

Increasingly, though, my experiences with bears have been more indicative of their feast-or-famine lives. A mother and two tiny cubs stranded on a small ice floe, adrift and far from the pack ice to the north or land to the south. The cubs were mewing like forlorn lambs, hungry, confused. The mother could swim 60 miles even in these frozen waters, but the cubs seemed doomed.

On expedition in Greenland, camping out on a beach during an unseasonal melt of the ice within the world's longest fjord, a young male made a beeline towards our party from several miles away, then just kept coming. He was skinny and carried a scar on his snout – potentially a war wound from an encounter with an even bigger male. His hunger drove him to pursue us, even though we banged our pots and pans, discharged rifles into the air, and set off flash bangs. At one point he was no more than eight metres away from us, and clearly showing predatory intent.

My most memorable encounter, though, was in the bleak seas to the northeast of Svalbard, the most accessible and rewarding of Arctic destinations. Seabirds soared constantly about the decks, and the scenery was ever-changing, from sugar-frosted spiky peaks to glazed golf ball ice caps, from creaking, calving glacier fronts to mercury-smooth velvet seas at the edge of the pack ice. Svalbard beaches are covered with ancient fossils, mossy whalebones, and Siberian driftwood, and you might glimpse a rare Sabine's or Ross's gull in among the ivory gulls and snow buntings. Bullying glaucous gulls smashed smaller seabirds from the sky and plucked them up in front of us, while Arctic foxes sprinted in to steal the bounty.

Svalbard is the best place in the world to find polar bears, with far more bears living in the archipelago than people. The islands of Kong Karls and Kvitoya in the east of the archipelago are the most important denning sites for female bears in the whole world. The time in question, I was in a sea kayak, paddling solo, out among the icebergs, about a mile or so away from our main icebreaker vessel, when one of the team came on the radio. 'Steve, Steve! There's a polar bear out there, it's coming your way!'

I couldn't at this stage see the bear, but I was excited and started to prep my cameras, ready to get a shot if he made an appearance nearby.

'Can you see him yet?' Graham, the camera op, asked. 'He looks like he's pretty close.'

I strained my eyes in the direction Graham was describing, but couldn't see anything at all. Minutes went by. The silence was fierce. I suddenly felt a very long way away from the safety of our boat and my crew.

'Er, Steve.' The radio crackled into sound. 'I'm not sure if we should be worried about this ... it looks like he's putting the icebergs in between you and him. I think he might be stalking you.'

The cunning polar bear was using the exact same technique he would when going into stealth mode to stalk a seal. He was keeping his body low, swimming in between pack ice, then crawling on his belly, keeping hidden from me behind pressure ridges and snow mounds. Then he dove completely under a chunk of pack ice and disappeared. This is a classic move polar bears will use as they begin to hunt.

The waters were black and inky, and I had no idea if he'd pop up right under me. Time stood still. I felt utterly alone and completely terrified. Then a huge white head popped up from behind a sculpted bit of iceberg, just a stone's throw away from me. He was standing up, sniffing the air, checking out his options. Perhaps sensing that the stealth game was up, and that I knew he was there, he settled on a more nonchalant option. He swam out in front of my kayak, barely looking at me, the very picture of not being at all interested in me ... until he was behind me. And then with huge powerful strokes of his front feet, he swam for the back end of my kayak, trying for another surprise attack. I powered away as fast as possible, putting distance between us, and headed straight back for the icebreaker, heart thumping and very ready to get back on board.

Several days later, I was woken at two in the morning by shouts of 'Bear!' It was summer, the time of the midnight sun, so it never got dark. Instead an eerie lilac and washed-out pink light stained the landscape – it was light without shadow. It was also bitingly cold, even in the stillness.

Before us were the bird cliffs of Alkefjellet, covered with 120,000 nesting Brünnich's guillemots as well as kittiwakes and glaucous gulls. It was a formidable sight, sound and scent. The birds wheeled round the pinnacles and towers like a swarm of bees, chuckling and caa-ing in their multitudes. It was breathtaking. It was not, however, why the crew had woken us at 2am. Swimming along the base of the vertical cliffs was a large polar bear. It seemed to be looking for somewhere to clamber out of the water. The team scrambled to assemble the filming kit, while I watched through binoculars the strange scene unfolding before us.

Finally the bear found purchase, and hauled herself up and out of the water onto the rock. From her proportions – especially the stoutness of the rear quarters and less massive shoulders – we identified the bear as being female. She shook like a giant white dog, fresh out of chasing ducks in the village pond. Then our bear began to climb. The cliffs were mostly vertical, but she picked a route using ledges and natural features in the rock, ascending upwards without stopping for rest. Approaching the top of the cliffs, she came to a gully packed with ice. Just as a human ice-climber uses stabbing ice axes and crampons, our bear used her claws like picks to penetrate the ice. At one point, with her whole body vertical in the gully, she even lifted one leg balletically up to waist height, in a perfect simulacrum of a human climber bridging their feet out to a hold.

Once up and out of the ice, our bear tiptoed along precipitous ledges, there to eat the eggs, chicks and possibly even slumbering adult guillemots. It was unexpected to see a giant predator so nimble and adept, but also sad to see her clearly so desperate for food that she would risk her life for a few eggs.

The polar bear is currently considered 'Vulnerable' on the IUCN* red list, and is known by Inuit people as Nanuq, which translates as 'the ever-wandering one'. They are an icon of the north, with a lonely life of starvation interspersed with occasional gluttony, in an environment that could not be more alien to our own. Yet respect and awe has not prevented as many as 300,000 polar bears being killed between the beginning of commercial hunting in the 1700s and 1973, when the Agreement on the Conservation of Polar Bears was put in place. This represents an average of around 1,100 animals per year and is responsible for the critical decline of the species, bringing them to the very precipice of extinction. What is even less known is that around the same number are still being hunted and killed each year, despite polar bears being officially protected.

Many of these bears are allocated 'harvests' for subsistence hunting, practised by native Arctic peoples. I've been lucky enough to spend a good deal of time living among the peoples of several Arctic regions, seeing something of their heritage and culture, and learning much from their experiential records of the realities of climate change. And while I have enormous respect for their traditions and customs, I have also witnessed practices I find uncomfortable. There are around 13.1 million people living within the circumpolar north, and very few have what could truly be called traditional lives. Most live in well-insulated homes, with high-speed satellite internet, hundreds of television channels, and modern clothing. They fish using motorised boats with fish-finders and satellite-enabled weather forecasting, and through the cold winters they rely on imported foods from stores that are on occasion stocked with exotic fresh fruit and vegetables. And why shouldn't they? They are custodians of a

---

* International Union for Conservation of Nature.

noble and ancient culture and still live lives with daily challenges I can barely contemplate.

Where it all starts to become a bit uncomfortable is when it comes to the hunting of endangered species. The Iñupiat have annual quotas to hunt unicorn-horned narwhals, white beluga whales and 100-tonne bowhead whales. Though this is classified as a traditional subsistence industry, hunters don't use hand-thrown harpoons and row boats as their ancestors did. Instead, they often use plane and helicopter spotters, then shoot whales from motor boats using explosive harpoons and automatic weapons. In the Nunavut region of Arctic Canada (where I've done several expeditions), local people enter a lottery to gain 'tags' to allow them to shoot a polar bear, worth as much as $50,000. However, those tags are being auctioned off to foreign hunters. Just type 'polar bear hunt' into Google and you'll find outfitters that can hook you up to pay a hefty fee, then go out and shoot yourself a bear. This constitutes trophy hunting, which is clearly in breach of the 1973 accord.

But while it is easy to get people to care about the travails of this icon of the frozen world, obvious physical dangers like hunting are not the greatest threat to their future. Instead, it is the broader changes to the habitat they call home that are their most bitter foes.

. . .

Like the deep, the cryosphere is a biome where most humans rarely tread. We can think of them and their wildlife as being remote, hardy and divorced from our own experience. Out of sight, and out of mind. Science, however, is just now starting to understand how interconnected our planet's environments are, especially under the insidious influence of climate change. The future of the leopard seal, the great ice bear and all the other animals that call the cryosphere home is on the line. We need to understand that if we compromise the health of their habitats, we will all suffer.

In biblical times people would wander the desert to try to find themselves. It's my belief that in the future, they will go to the Arctic. We'll call it polar perspective, a pilgrimage to this savagely beautiful place for existential revelations, where the human condition is put into the starkest of focus. The concerns of our fragile modern world seem inconsequential before a polar squall. Our narcissistic self-importance is laughable in a place where we are clearly so powerless before nature. The challenges of our crypto- and cyber-obsessed society fade to dust when you're chilled to the bone – the comforts of our pampered existence pale in comparison to the warm food or fire that ends your exposure to the elements.

We as individuals are nothing before the great power of the cryosphere, but we as a species have been so insidious that we threaten the very existence of the frozen zone. As the pack ice shrinks, and the glaciers crumble, there are implications for everyone, wherever they live.

Climate change has been most real for me in the Arctic, because this critical place, where we feel most alive, is changing faster and with more devastating violence than anywhere else on the globe. Temperatures are increasing twice – potentially even four times – as fast as anywhere else.

On Baffin Island I met local Iñupiat peoples who have an oral tradition dating back thousands of years, and an intimate relationship with their land. Their calendar is based around the spawning of the charr, the blooming of the sourdock, fireweed and cotton grass, and crucially the break-up of the sea ice. These timeless landmark occasions had been the same for millennia and dictate every waking hour for Arctic peoples.

The Iñupiat are utterly bewildered that anyone in the outside world could doubt climate change. One friend took me on a tour around the bay in front of his house. 'Up there was a glacier until the late 1980s. The ice here is forming nearly a month too late, and breaking up a month too early. Polar bears, starving and desperate, come into the village dump to feed and we have to shoot them. We have fish in our seas we've never seen before, others that we used to rely on have not been seen for years, and there are now insects here we don't even have names for.'

I did my first programme on climate change in the Arctic in the year 2000, and my first full series on it in 2005. The irony is not lost on me that we did much of our travelling around the Arctic in helicopters; however, we did work with some of the world's finest climate scientists, and we did some real and exciting science. We bounced sonar off the bottom of glaciers to see how an increase in melt was speeding up their movement from both below and above. We went up into the permafrost, where the soil has been frozen solid for 2.5 million years, and found that as it was melting for the first time it was emitting vast amounts of methane, one of the most potent greenhouse gases. To prove it, you simply had to light a match, and whooomph!

In Alaska we compared photos of several glaciers to historical photos taken two decades previously. It was a struggle to even line the photos up as the ice had receded so far. On that same trip, a climate scientist friend showed me that in the northern tundra of Alaska, permafrost had sealed the soil since the last ice age. It was now soggy, defrosting mud, and the vegetation was starting to rot. We abseiled down into moulins, huge whirlpool plugholes in glaciers, where icy meltwater tumbles down into their guts. These moulins are getting bigger, form earlier in the year and have increased in frequency, and every one sends water to the bottom of the glacier, lubricating its movements, speeding its demise, ultimately increasing the amount that calves off at the end to become icebergs.

In Greenland we came across ancient bergs with ice as clear as diamonds. When you stuck a chunk into your whisky, it vaguely fizzed, as compressed microscopic air bubbles were released. Air that had been trapped in the ice for 100,000 years was breathing again. With 70 per cent of the world's freshwater sealed in glaciers, ice caps and permanent snow, it's clear that drastic melting will lead to global sea level rise.

Back in Alaska, I worked with a biologist who was studying the stunning wild mountain goats and Dall sheep. In the summer they are plagued by biting blackfly, and move up to the highest slopes to evade their attentions. With every 200-metre vertical ascent it gets a degree cooler, and the blackfly cannot follow. However, the increase in temperature has already meant the parasites are comfortable even on the highest peaks, and the goats simply have nowhere higher to go. The goats now wander round in incessant circles, unable to feed during the crucial time of summer plenty, and thus their breeding success plummets.

There are also very real challenges for the whales of the north. Some of my greatest ever wildlife encounters have been with humpbacks heading to the north to gluttonise and gargantuate on the summer surfeit of herring. It is often believed to be the finest wildlife spectacle on the planet. But the lives of the whales are tied to those periods of summer prosperity, and their prey moves in accordance with vast complex currents, driven by the ever-changing salinity and temperature of their seas.

In 2020, just south of the Arctic Circle in Iceland, we worked with perplexed cetacean specialists who reported the normal food fish had been replaced by explosions of mackerel, which the whales don't seem to eat. As a result, their whale sightings plummeted.

Even greater challenges face the true Arctic whales, the bowhead, beluga and narwhal. As the sea ice fragments, it will allow more boat traffic into their once sacred seas. Hunters will be able to reach them, there will be more oil spills, and novel pathogens to which they have no resistance will be introduced. Unpredictability in ice formation will lead to more animals getting trapped in the ice, and their one predator other than man – the orca – will have unfettered access to them and their calves.

We have such an anthropocentric view of the world. We see animals like the polar bear that can survive in seas at –1.8°C and under Arctic ice as being tough and rugged, because it would be such a challenge for us. But they are a result of millions of years of natural selection that means these wild waters are perfect for them. Existing in such a narrow niche leaves them vulnerable as it changes. The animals that may seem to us the most robust are in some ways the most fragile. The polar bear may be the epitome of ferocity, of resolute solo stoicism, but in reality it is a wanderer on a narrow trail, with a precipitous drop should its course falter.

CHAPTER FIVE

The Bounty

The ocean asks for nothing but those who stand by her shores gradually attune themselves to her rhythm.

Charles Dickens

We've been at sea for two days, but all days are not created equal. There are few longer days than those spent on a small wooden boat, bouncing over whitecaps with a force that seems to want to separate your spine from your pelvis. The team are red raw from sun and wind burn, flayed by the Yucatan summer scorch. And then I see it. Way out at the horizon, perhaps two miles away, what looks like a swarm of bees down close to the surface.

'Baitball!' I shout, pointing like a madman out to the swirl of predatory frigate birds, swooping down onto baitfish. It takes us 15 long minutes to reach the melee, each of us begging Poseidon not to let the feeding festival come to an end or for a sport fishing boat to find it first.

Finally, we are within range of the birds, perhaps a hundred great black pterodactyls swooping on piratical wings to snatch the silvery sardines from the surface. No one has yet seen what it is that's driving the fish up from below, but we all have an idea.

Cameraman Simon and I will be diving together, with no one else in the water, meaning one less body to get speared or drowned. There will be no time for scuba or rebreathers; we will have to drop over the side with just the breath in our lungs.

The second we splash in, we are in the midst of a tornado of quicksilver. The baitball is a shoal of platinum-sided fish each but a foot long, driven together into a shimmering sphere for protection. The trigger for the shoal to ascend to the surface is about 40 of the world's fastest fish. The sailfish is a super-charged assassin, long and slender, their snouts wielding a mighty rapier that they sweep through the baitball to separate out individuals. Their flanks change colour as they scythe through the water at speeds of up to 50mph, vibrant aquamarine, stripes of silver and neon. They hang off at a distance from the shoal, barely twitching to keep pace with it, before propelling themselves in at what seem like impossible speeds.

This is the most physical dive encounter I've ever had; powering along with all our might to keep up with the action, hauling ourselves back on board, then plunging back into the water. It goes on for nearly an hour. As the baitball diminishes, the fish start looking for safety anywhere they can find it. A small splinter group comes in between me and my underwater camera, and a sailfish lance comes in at breakneck speed to joust them out again, risking turning me into a human kebab.

Then there is but one lone fish left, twitching injured in front of Simon's lens. A sailfish swoops in with ominous ease, snatches it, and then they are gone. All that remains of this *violent feast is silvery scales drifting down through the water like falling snowflakes.*

. . .

There are more than 33,000 known species of fish on our planet, with new species being described all the time. That's more than all the mammals, birds, reptiles and amphibians put together. They come in a bewildering array of shapes and sizes, from the mighty whale shark that can be the size of a bus to tiny cichlids and gobies smaller than your little fingernail.

The array of shapes found in fish reflect their central challenge: how to move around in a medium 800 times denser than air. The abundance of different fish forms and flowing tail fins is testament to how evolution has attempted to conquer motion in the stubborn matrix of water.

The kings and queens of streamlining underwater are the billfishes, particularly sailfish, which are so hydrodynamic and possessed of such explosive musculature that they can reach speeds of 50mph. The tuna is perhaps even more impressive, able to maintain its explosive speeds for sustained periods of time. Many billfishes can also heat their eyes and some organs to well above the ambient temperature of the water around them by running blood that has been warmed in exertion alongside cooler blood. This potentially makes them ten times more effective at reacting to the actions of their prey.

The symbol of Darwinian evolution is often shown as a fish with legs, simultaneously a snub to the ichthys, the Christian fish symbol, and a reference to modern terrestrial life being ultimately evolved from fish. This is a pretty big leap, and has more holes than the average sieve, but the idea of all life evolving from fish does embody certain truths. Rather than saying that we and other mammals evolved from fish, it is more accurate to say that we share a common ancestor, with evolutionary biologists highlighting 'flaws' that we inherited from these swimming ancestors.* The closest thing that science has found in the fossil record to back this up is approximately 405 million years old; a fossil which shares characteristics of both fish and land vertebrates. After this there were likely two main branches, one of which became modern ray-finned fish, and the other of which evolved into the group that contains lungfishes and tetrapod vertebrates.

What is unquestionable is that fish have been around for a very long time. Five hundred and thirty million years ago, during the Cambrian explosion, the first chordates developed skulls and proto-spinal cords. Then came the jawless fishes or agnathans, eel-like fish called conodonts, armoured ostracoderms, then their descendants the placoderms. Most of these ancestral forms are long gone, but a few jawless fish linger today, in the form of the deep-sea hagfish, which feast on whale falls, and lampreys, some of which can still be found in British brooks.

Not only are new species of fish being discovered all the time, particularly from our deep seas, but there are also old species being rediscovered too. In 1938 in a fish market in South Africa, ichthyologist Dr Marjorie Courtenay-Latimer saw and eagerly purchased a strange, ugly fish. She recognised that the beast before her had gone extinct 66 million years before, around the time of the demise of the dinosaurs. Finding a coelacanth here

---

* Often quoted as evidence against intelligent design is the recurrent laryngeal nerve, which branches off the vagus nerve. As fish don't have necks, this nerve evolved to go directly from the base of the brain, around the arteries and on to the larynx. When terrestrial mammals evolved from this common ancestor, they couldn't 'rewrite' this part of their biology as their bodies changed. In the modern-day giraffe, this nerve now goes all the way up and down its neck to get to the same place; a nerve that should be just centimetres long is instead five metres long.

was like finding a velociraptor in a rural farmer's market. They were a part of that first group that diverged from ray-finned fishes 405 million years ago.

The great period of time since the first fish arose has led to the development of some true oddballs. There's the flamboyance of the weedy and leafy sea dragons, huge stretched seahorses with intricate tendrils and baubles trailing from their impossible forms. Diving alongside these bewitching oddities causes a strange optical illusion: shine your white torchlight on them and they glow with brilliant colours – purple, neon blue and yellow – but avert your beam and the colours vanish, the sea dragons disappearing into the kelp like dappled green phantoms.

One of the oddest fish is surely the giant sunfish or *Mola mola*.* These giants are record breakers, with the biggest disparity in size between their tiny egg (the size of a pea) and adult animal (which can be six metres across). If a human baby exhibited the same growth as a sunfish, an adult person would be the size of the Royal Albert Hall. In Nusa Penida off the coast of Bali, the *Mola mola* will come up from the depths with the cold water in order to be groomed by cleaner fish. Yellow and black butterflyfish attend to its needs, swimming right into its open mouth and picking at nagging loose scales and parasites.†

Finding *Mola mola* here involves a logistically challenging dive, which I've attempted several times but only managed to make work once, in 2018. The encounter only tends to happen deep and very early in the morning, when the cold water layer extends upwards towards the bottom of the reef. It's truly remarkable quite how much these giant fish are bound to the

---

* The scientific name means 'millstone', and refers to their circular form, which resembles an old windmill grinding stone.
† While filming off the coast of San Diego, a sunfish rolled onto its side at the surface, allowing both a nearby albatross and me to pick the lice off its skin!

cold thermocline. You can see *Mola mola* as tiny dots off in the distance, 30-odd metres below you, and then a freezing chill bites through your wetsuit as the thermocline shifts, and the next thing you know a giant sunfish is hovering right in front of you, being groomed by an entourage of barberfish and cleaner wrasse.

As these dives took place at 5am, it left an awful lot of the day free to explore. We'd done most of the obvious and well-known dive sites, so we hit on an unusual plan. We'd just dive somewhere that was not an established dive site and see what an ordinary, unprotected, unmanaged, uncleaned Bali reef looked like. The results will stay with me always.

Put-putting around the southeastern coast of Bali, we saw a nice-looking cove with a few guesthouses, and a long pier jutting out into the bay. As with everywhere on the paradise isle, building work was underway to feed the insatiable desire for more and more hotels, and regular boats were chugging into the side of the pier to deliver planks and boxes of tiles and other building materials. The sea was clear enough for us to see the bottom, and the dark shapes of coral heads were clearly visible, with occasional flickers of fish seething over the sand.

Not long after submerging, we could see there was a functioning coral reef here, although it looked a bit like a sunken rubbish dump, covered with every bit of detritus you can imagine, from sandwich boxes, sanitary towels and crisp packets to a discarded potty and empty-eyed dolls sat staring grimly from the coral. Most distressing though was the fishing gear. The coral heads were criss-crossed with monofilament line, stretching in straight strips for hundreds of metres, curled together into tangles where it had been cut free once snagged. Running for at least 100 metres was also a so-called 'ghost net', a fishing net that had become snagged on the coral and then dumped there. These nets are designed to catch fish, and they carry on doing that job long after they've been dumped.

Our first victim was a long elongated trumpetfish, snagged and battling gamely to get free. Amazingly, hovering alongside it was its mate, itself totally free, but unwilling to swim away from its fettered partner. This was a scenario we saw replicated with half a dozen different species: for example, two scorpionfish side by side with the larger (presumably female) caught and thrashing against a line she could never in a thousand years break. The maxim of keeping your distance and not intervening if a wild animal is in danger does not stand when their challenges are of human origin, and the team set about cutting this individual loose. The second she was free, she fell in place alongside her mate, though as bottom-dwelling sedentary fish, they didn't move far from the net, which would surely have caught them again. I found it surprisingly affecting. We incorrectly think of fish as being little more than insects in terms of sentience, and yet here they were displaying what appeared to be loyalty, partnership, even love.

Perhaps the most unusual species we freed from the ghost net were two giant frogfish, one with its hand-like flippers caught in the line. This is truly one of the oddest fish on the Indo-Pacific reef, and remarkably similar to the unrelated deep-sea anglerfish. Fat as a flat football, they are phenomenally camouflaged and able to change colour to match their environment. They're not much for swimming, with a tiny tail that looks comical alongside the corpulent body. Instead, they walk along the bottom on their hand-like pectoral fins. At the top of their heads is a modified dorsal fin with a fleshy lure on top. They can sweep it over the front of their heads, in order to attract small swimming creatures towards the suction action of their cavernous mouths. They're often found in pairs, with the female being slightly larger and the two often being totally different colours. In this case one was mottled brown and dark red and the other was bright buttercup yellow.

When we released the trapped frogfish from the net, it swam up onto Mark's camera, seemingly trying to hide behind his lens. When the other followed, we were rewarded with the frankly ludicrous sight of two swollen frogfish standing on a camera half their size, as if trying to say, 'You can't see me … right?'

As we continued to swim the length of the ghost net, we found stingrays, flounder, cuttlefish and others that had not been so lucky. Tangled into the net, they'd starved or suffocated and then been picked at by other fish, which had also then been snagged. We pulled up 20 or 30 metres of net, then returned for a second dive to try to pull up as much more as we could, but it was a literal drop in the ocean. Clownfish in their anemone homes flitted about beneath the net, coral heads hundreds of years old were shredded and entangled, and cutting them loose in some cases was as destructive as leaving the nets there.

Just the day before, I'd spent a fun day with a squadron of local kids, part of the Plastic Clever initiative set up by inspirational youngsters Melati and Isabel Wijsen. The girls had secured an audience with the Indonesian prime minister by going on hunger strike, spoken at the UN, and done a global TED Talk. They were genuinely some of the most inspirational young people you could ever meet, very much at the vanguard of a youth movement whose aim is to ban the sale of plastic bags and bottles completely and make Bali free of single-use plastic. And their efforts are needed. Indonesia disposes of 6.8 million tonnes of plastic rubbish every year, with that amount increasing by 5 per cent a year. That is nothing like as much per capita as the UK, for example; however, as an archipelago of 18,000 mostly small islands, more of it ends up in the sea than from any other nation.

With the enormity of the challenge ahead of them, Plastic Clever are going to need to recruit a huge number of Indonesians to their cause if they are to avoid swamping some of the world's most special seas in single-use soda bottles. Studies in 2023 showed that there are now 171 trillion pieces of plastic in our seas, and this could potentially triple by 2040. Mind-boggling and horrifying statistics indeed.

. . .

Standing on the bow of a boat surging out over the Pacific, the expanse before you seems to have no end. In every direction to the horizon, the view is unchanging, unwavering, eternal. Though the water itself is yielding, the lashing of its incessant swell is the most destructive and overwhelming force on earth. It is the power of an entity that doesn't have to rush; it can take its time to move mountains.

The blue of the ocean is so intense that it envelops even the sun; red light that has travelled 152 million kilometres from our life-defining star is gone in a few metres of blue, blue, blue. It seems to suck the soul from your bones. There are places where you can row or even sail for months and never see land.

It's not surprising then that for millennia people have felt the bounty of the oceans must be infinite, that you could take what you want and their resources would never run out. But fishermen and governments have known for a dozen decades that this is not true. The great lie has been to keep the ocean's limits secret. To pretend we can keep strip-mining the depths and like a magic trick the fish will just regenerate. Much like the world's petrochemical giants managing to sell us the lie that climate change is nothing to worry about, the giant fishing companies have pressured our governments to keep their secret safe. That secret is that nothing about how we use the sea is currently sustainable. We can and will empty our seas without immediate global action.

Statistics show that we humans catch between 1 and 2.7 trillion fish a year, or 100 million tonnes. It's probably the most repeated fact in modern marine biology that by 2050 there will be more plastic in the ocean than fish. Less quoted, but in some ways even more bleak, is that by 2050 large predatory fish will be gone completely. Over the last hundred years, the populations of these fish have depleted by two-thirds. By 2050 there may be many small fish left, but these are not necessarily the ones we choose to eat. We may take around 100 million tonnes of fish from the oceans every year, with that amount growing, but catching those fish is getting harder and harder.

To me, the most head-spinning evidence of our consumption is found in Tsukiji Market in Tokyo, the biggest fish and seafood market in the world. It sells 480 different species of seafood from 60 countries, from sea urchins to caviar, abalone to zebra mussels, and tuna to whale-meat (killed for 'scientific research', yet curiously still on sale here for food). Strolling round the lanes among the 65,000 employees, you'll see live seafood slopping around in buckets, langoustines making a break for freedom from their crates, and octopus arms lashing across their tanks.

About 3,000 frozen bluefin tuna are sold every day. On my first visit in the late 1990s I watched an auction where one bluefin tuna went for a jaw-dropping £80,000. Which is nothing. In 2021 a single fish went for £2.4 million. The increasing rareness of the bluefin tuna means these fish command higher and higher prices, and are a publicity-generating status symbol for the sushi chain that manages to acquire the most perfect, most expensive specimen. That kind of cash does not exactly inspire a rational and considered approach to sustainability. Pacific bluefin tuna have lost 96 per cent of their numbers, Atlantic bluefin is considered even more endangered, and 90 per cent of fish on sale at Tsukiji are too young to have bred.

However, the thing that really shocks is the sheer scale. Every year more than 700,000 tonnes of seafood is handled here; around 6 billion US dollars' worth. I think every visitor to Tokyo should visit Tsukiji as a cautionary tale. It is a city-sized aquatic abattoir, which brings home quite how much we are strip-mining our seas.

So what are the solutions? Well, we have to stop using fish as food for agriculture and aquaculture, and enforce our quotas on specific species that cannot take the hammering of intensive fishing. We also have to ensure we strategically protect as much of our oceans as possible. Perhaps the most heartening conservation news of my lifetime was the signing of the High Seas Treaty in early 2023. After a decade of negotiations and under the auspices of the UN and the IUCN, member states signed up to protecting as much as 30 per cent of our high seas by 2030. It's a huge order to police and patrol this much sea, and any marine protected area will inevitably be an attractive fishing site for the unscrupulous, but it is a heartening beginning and could be the start of the saving of our seas.

. . .

Our oceans possess the space and volume of life that we need to solve our planet's problems. For example, one of the biggest potential sinks for greenhouse gases is so-called 'blue carbon'. Plants such as mangroves and seagrasses are some of the planet's most underrated ecological buffers. In 2021, UNESCO published a report suggesting that although only 1 per cent of our oceans are made up of seagrass, they form an unparalleled carbon sink. Shark Bay in Australia is the largest documented seagrass bed in the world, and the Sundarbans in the Bay of Bengal are the largest mangroves. Scientists suggest these environments alone are able to suck in as much as 10 per cent of our greenhouse gas emissions. These blue carbon ecosystems are one of the most potent ways of mitigating anthropogenic climate change. Seagrasses photosynthesise just like terrestrial plants, and a large part of their biomass ends up in the soil on the seabed; this carbon is therefore locked up until those seabeds are destroyed. The soil is also extremely oxygen poor, which inhibits decomposition, and the plants' roots stabilise it, binding it together. The stems, meanwhile, slow currents, mitigating the worst effects of waves and storm events. This is also true of mangroves and salt marshes. In the UK alone we've lost 90 per cent of our seagrass meadows. Half of that has been lost since the turn of the Millennium and we now only have around 32 square miles; an area the size of Newcastle.

As we destroy these vital but unappreciated ecosystems, we are cutting ourselves, then pouring vinegar into the wounds.

Studland Bay in Dorset, England, is one of the UK's best-known remaining seagrass meadows. Diving in among the seagrass, you encounter tiny versions of the bigger fish you'll find further out to sea, as this is their nursery. There are black gobies with bulging cheeks and puckered lips, tiny cuttlefish, squid, and pipefish hiding in among the rhizomes of the seagrass, Medusa-like snakelocks anemones fastened to the stems. The biggest treasure to be found here though is the UK's two species of native seahorse.

In Studland, the populations of seahorses are in constant flux. During the aftermath of the first COVID lockdown, I was given a special permit to dive here with Neil Garrick-Maidment of the Seahorse Trust. It took us no more than five minutes to find our first seahorses, twisted around the stems of the seagrass. We soon found a pregnant male, his brood pouch engorged with tiny baby seahorses. The question as to why the males carry the young is intriguing. The prevailing hypothesis is that as seahorse fry are so small, slow-moving and vulnerable, simply jettisoning them as most fish do would result in a sky-high mortality rate. (It's extremely high for all fish, but in this case the odds would be untenable.) The male carrying them for a precious month gives them the time to grow into miniature seahorses that have a slightly better chance of reaching maturity. The reason the female doesn't carry them is so that she can get back to feeding, and then nourishing a new set of eggs, ready to start the process again as soon as possible.

No more than a few weeks after our dive, as business-as-usual tourist season recommenced, speedboats ploughed through the seagrass, ripping up the meadows and the soil with their anchors and engines. Once the substrates were disturbed and the seagrass roots torn apart, they washed away, leaving bare sand. Studland Bay is on paper a marine reserve, but there is no one to enforce that protection. When Neil next dived in the bay, he didn't find a single living seahorse.

In 2023, the United Kingdom tried to revamp its pointless, toothless marine reserves by declaring three genuine no-take zones around our coasts. However, in a stunning display of short-term political expediency over longer-term gain, the original five zones were reduced, as it was felt they would compromise fisheries too much.

Perhaps the most galling thing about the situation at British marine reserves is the fact that genuine no-take zone reserves are the biggest silver bullet in all conservation. Perhaps the starkest evidence of this is found in California's Monterey Bay, which has been an epicentre for fisheries for 240 years, and as such has had every single one of its marine inhabitants brutally and fatally overexploited.

One of the first wild creatures to be exterminated was the sea otter. As a marine mammal that is not insulated by blubber like whales and seals, sea otters possess the densest fur on the planet, consisting of 100,000 hairs per square centimetre. It's an essential adaptation in waters that are blessed by cold upwellings of nutrient-rich waters, the very basis of the flourishing food chain. But their fur coats proved to be their downfall, as waves of both legal and illicit hunters ravaged these shores in search of the valuable pelts. In the late 1700s they were seen as the royal fur in China, and in the 1800s a single pelt could raise $100. Not surprisingly the last sea otter disappeared from Monterey before the end of that century.

They were not the last animal in the area to perish before human avarice. In 1854, a whale was worth two pounds of pure gold, and hence they too fell, first to spears then to the explosive harpoons of the industrial whalers. Where once it was said you could walk across the bay on the backs of the plentiful whales, the business eventually collapsed as there were simply no whales left to catch. In the 1800s, abalone brought the Chinese here, and seabird eggs were food for the prospectors of the Gold Rush. Then from 1910 to 1940, Monterey Bay was home to the biggest canned fishery in the world, as immortalised in John Steinbeck's *Cannery Row*. Seals and sea lions were clubbed out of existence, industrial pollution clogged the bay, and the waters were strewn with the guts of a billion sardines. It was as though every animal was targeted until it was functionally extinct, and then the humans would move on to something new.

But then the tide turned. Many give credit to one of the greatest unsung conservationists of all time. Julia Platt was the headstrong local who became mayor of Pacific Grove, a city in Monterey County, in the 1930s (when a female mayor was unheard of), and she was the first to have the courage to take on the curmudgeonly might of the fisheries. She not only fought for regulation but for complete no-take zones, and she was part of a sea-change in attitudes, seeing environments and animals as having value above and beyond economics. The refuge outside the Hopkins Marine Station still bears her name.

The real triumph of her ethos happened quietly sometime in 1938. Without fanfare, somewhere down the coast, a sea otter returned, probably from the north, where their habitat still thrived. And then hundreds, and then thousands. Obviously this was a cause for celebration. Few animals are cooed over more by the general public than cute and cuddly sea otters (despite the fact that, as the largest member of the weasel family, sea otters are the size of Alsatians with teeth to match and will occasionally kill the local harbour seals). But nobody could have predicted what happened next.

Sea otters dive to the seabed, snatching up crabs, abalone and sea urchins, before bringing them to the surface and smashing them open on their stomachs – often with the aid of an anvil stone they keep stored under their armpit. When the sea otters were decimated, their prey sources exploded and devoured the kelp, which in turn disappeared. But sea otters have to eat a quarter of their substantial bulk every day. When the sea otters returned, so did kelp forests that had not been seen in Monterey for over a century. With urchin and abalone numbers kept in check, a whole environment bloomed.

Composed of the fastest-growing plant (strictly speaking an algae) on earth, these dense, dark forest canopies are exactly like a submarine rainforest – green caves illuminated by sunbeams that cut down into the bustling jungle beneath. The tendrils and wave-lashed wisps of weed are home to a bustling multitude, from tiny fish right up to giant pouty-mouthed bass. These forests became nurseries for the fish that were to restock the bay. Harbour seals and California sea lions returned in their droves to feast on the bounty, as did pelicans, gulls, sooty shearwaters and cormorants, in numbers that would be declared plague proportions ... if there wasn't clearly enough food to feed them. And then, finally, humpback, sei, fin and eventually blue whales came back. They found food in vast shoals of baitfish, as well as a deeper layer of crustacean krill that seemed everlasting.

I first filmed at Monterey Bay in the year 2000, and found it a breath of fresh air in the clamour of ocean catastrophe predictions, which were already omnipresent in conservation back then. Fifteen years later I returned with a grand BBC project, which aimed to show the wonders of Monterey Bay live to hundreds of millions of viewers around the world. At night we dived among gargantuan shoals of market squid, which come here in huge numbers to spawn. In a single day we recorded seven species of whale and dolphins, including the largest animals ever to grace our planet, mighty fin and blue whales. White-sided dolphins lunged and snorted in the bow wave of our scientific vessel as we raced along. There were also curious melon-headed Risso's dolphins, their flanks scarred from battles with deep-sea squid (and each other), and humpbacks that had journeyed across the vastness of the Pacific to feed their starving calves in the safety of the bay. I hope that Mayor Julia Platt would have been proud ...

The story of Monterey gives me great hope. If a once functionally extinct environment can bounce back to become one of the world's great marine sanctuaries, and that can happen in a matter of a few decades, we can turn the tide for all of our fish and wildlife. Proper no-take zones work, and the results can be awe-inspiring. This to me is the answer. Science can quantify the damage we're doing and explain the value of environments that we might not otherwise care about. And science can come up with plans to save our seas and find ways to put them into practice. We're not going to ban fishing any time soon, and it will be a fearsome battle to try to enforce the new treaty in the wild, lawless frontier of the high seas. However, if we can properly protect some of our most strategically valuable coastal waters, then we can have a huge knock-on effect on the overall health of our seas. It just takes an act of will from people who care about their patch. That, and a few heroes willing to stand up and shout about what they believe in.

CHAPTER SIX

Survivors

Sharks have everything a scientist dreams of. They're
beautiful – God, how beautiful they are! They're like
an impossibly perfect piece of machinery. They're as
graceful as any bird. They're as mysterious as any animal
on earth. No one knows for sure how long they live or
what impulses – except for hunger – they respond to.
There are more than 250 species of shark, and
every one is different from every other one.

Peter Benchley, *Jaws*

*The darkness beckons us in. Smoke cloud shoals of colourful fish shift
and ripple in unison like a single living organism. Pouty-lipped
groupers hang motionless as if sleeping, their sides pocked with dark
blotches like a child's potato prints. My torch beam slices through
the gloom like a death ray. The light gleams copper on sleek skin.
Lazily, the shark glides out of the cave and into the light. This shark*

goes by many names — grey nurse, sand tiger, ragged tooth — and for many, it is the very epitome of a nightmare shark. Hundreds of curved fishhook teeth spill from their jaws like something from a monster movie. Their eyes are small and staring, with a tiny dark central pupil. There is no glimmer of the personality that I've seen from other species. They are known to cannibalise their siblings while still inside their mothers. And (this next part is difficult to quantify), they seem more 'fish-like' than other sharks. Perhaps it's something about their tail and body shape. Or because they hang motionless midwater among the shoaling cave residents, rather than constantly swimming like many other shark species, or lying on the bottom like carpet sharks do. This strange feat is achieved by gulping air at the surface to offset their buoyancy, and functions like a bony fish's swim bladder.

As my eyes become accustomed to the gloom, I cautiously fin my way into the cave. Hanging ominously there are another 30 shark shapes, like oversized evil salmon, flank to flank. Even the most ardent shark lover would have to concede it is a sinister sight. I fight the urge to turn back to the light, to flee the hovering horde of demon spawn before me. But as I push on into the cave, the sharks melt away into the shadows and are gone.

· · ·

I remember with crystal clarity the first time I saw a shark. I was nine years old, and my family and I were in Tioman Island in Malaysia, which in 1982 was the closest thing to paradise imaginable. My days were spent obsessively snorkelling around the near shore reefs, pondering the parrotfish, and diving down to gaze at the sponges and spear mantids. One day I noticed a familiar silhouette coming out of the blue swirl towards me. My first ever shark! My heart leapt — not with fear, but with excitement.

Forgetting all the rules I'd learned about wildlife, I swam after it, hoping to get a closer glimpse and extend my encounter with this ancient miracle. But after a few minutes it became clear I didn't need to chase the shark. Far from it. In fact, if anything it was getting closer. And swimming round and round me in ever-decreasing circles.

With my mind filled with images of tanned surfers punching sharks in the face and Jacques Cousteau riding on the dorsal fin of a behemoth shark, I did what any self-respecting nine-year-old adventurer would do. I scrambled up onto a rock and sat there for two hours until sunburn forced me back into the water. I swam back to shore faster than Michael Phelps, and didn't get back in the sea for the rest of the week.

In my fever dream, my shark was megalodon-sized and could have eaten me whole without needing to chomp. In reality it was a black-tip reef shark, and about the size of your average haddock, less danger to me than a house rabbit. There was a lesson to be learned right there about sharks – perceived risk and actual danger are rarely the same thing. However, this marked the beginning of a fascination bordering on obsession with the lords of the sea, one that has seen me spend several thousand hours underwater with them in every ocean – bulls, cows, tigers, leopard and zebras, bronzies and silkies, threshers, makos, oceanic whitetips and even great whites without the sanctuary of a cage. I've hurdled a great hammerhead while being watched on live television by 5 million people on the BAFTA-award-winning *Blue Planet Live* in 2019,* used UV lights in midnight kelp beds to illuminate neon glow-in-the-

---

* I was standing on the seabed and a male hammer swam straight at me along the bottom. I leapt upwards through the water, and it swam under my feet, much to the delight of people on Twitter.

dark cat sharks,* and been sucked into the mouth of a feeding whale shark.

For years I chanted the mantra that sharks are hopelessly misunderstood and basically harmless, before having to acknowledge my own misconceptions and re-evaluate my respect. Even now, my understanding of what a shark is and does can be radically changed on just a single dive. Anyone who thinks sharks are malevolent monsters is completely clueless, but likewise anyone who claims to know what a shark is going to do next is kidding themselves.

Perhaps no animal in our ocean incurs such extremes of wrath and rhapsody, terror and passion, as the shark. Scientific analysis has revealed that most people are terrified of sharks,† with even the vague possibility of their presence being one of the most common reasons for a fear of the sea. However, very few of us ever physically encounter them, with most of us only experiencing sharks through the media. Further analysis has shown that the majority of media representations depict attacks and use demonising language, and only around 10 per cent of content refers to threats to their conservation. Not surprising then that even today they remain an animal that we need to learn to love.

There are more than 450 extant species of shark, ranging from the biggest fish in our seas, the whale shark (which can be 20 metres long and weigh 40 tonnes), to the tiny dwarf lanternshark, which is the length of a standard ruler when fully grown.

---

* Initially bioluminescence was only known from a few select species such as kitefin and lanternsharks. They create their own light, which may aid countershading camouflage in lightless deep seas. We now know that many species – including common cat sharks in British seas – will fluoresce under UV light, probably enabling them to locate other members of their own species during breeding times.

† In a 2021 analysis, all 400 participants recorded their perception of sharks between 5.9 and 6.5 out of 10 on the 'fearfulness scale'.

Their array of super senses is more Space Age than primeval, and makes them even now the ultimate predator of our seas.

A classic shark would be from the family Carcharhinidae, known as requiem sharks,* a broad group that contains 60 familiar species, from bulls and tigers to reef sharks. They generally have a fusiform shape that is spindle-like, tapering at either end and highly hydrodynamic. The colouration is classic counter-shading, the most common marine camouflage. As we saw earlier with penguins, this consists of a darker dorsal (upper) surface and a light belly. The idea is that if you are above a shark looking down, then its dark back blends into the deep sea beneath. If, however, you are beneath it looking up, its creamy belly melts into the light of the sky and sun above. It sounds too simple to be true, but just about every good-sized ocean predator from the penguin to the great whale has it.

But sharks' senses are key. Said to have better night vision than a cat, their smell is probably their most acute and highly developed super sense. They can perceive certain substances such as the amino acid serine (present in fish blood) 100 million times more acutely than humans, even a single drop in an area of water the size of an Olympic swimming pool. It's often said that sharks smell in stereo, able to differentiate between the time when a scent molecule enters one nare (or nostril) and the other, and therefore to detect directionality of odiferous substances in the water.

Although they have no visible external ears, they have internal ears embedded in the skull cartilage and linked to the outside via two pores that terminate on the top of their head. These organs have a vital role in balance and orientation, but sound is important to sharks too. They respond most enthusiastically to pulsating vibrating sounds in the 25–600Hz range.

---

* The name may be from the French for shark, *requin*, or from the requiem mass for the dead.

The two other shark super senses are surely their most exciting. The first super sense is that generated by the lateral line and pit organs. The line branches from the head and then courses down the side of the animal. In many fish it is clearly visible. It contains sensory cells swamped in jelly which transmit pressure changes and vibrations in the water outside. This and the similar pit cells enable the shark to sense a fish that has swum past and is long gone, simply by the wake they leave behind. The second super sense is generated by what looks like a peppering of highly squeezable blackheads concentrated around their snouts, but which are actually part of an acute electro-sensory organ called the ampullae of Lorenzini, so called for the Italian physician who first accurately described them. Probably having evolved from the lateral line, these organs in the snout detect weak electrical fields and can even pick up the zappy signatures of animals that are completely hidden, concealing their bodies but not the minute fields given off by their beating hearts or the processes of their brains or circulatory systems.

The ampullae of Lorenzini are incredibly sensitive but only work over a very small distance – in fact, most sharks will touch a subject in order to process its electrical field. When diving off the coast of Cuba we dangled a sensitive hydrophone into the water alongside our boat above. The silky sharks that circled around us would make a point of swimming straight upwards to touch the mic with their snouts, undoubtedly sensing the weak electrical pulses emanating from it. And oceanic whitetips in open water will swim into our cameras, nuzzling the glass with their sensitive snouts, as if trying to figure out what on earth the electric buzz they're experiencing might be.

One side effect of this weird super sense is that it can be used to hypnotise certain kinds of shark. Hyperstimulation of the ampullae can lead to a shark simply falling into a catatonic state known as tonic immobility, in which they lie stupefied until seeming to remember where they are and shaking it off.*

. . .

The evolution of sharks is better understood than that of many other marine organisms, and for several good reasons. As successful creatures, at the apex of every food web they inhabit, they attract the attention of more than their share of biologists.

Sharks have been around for a long time. And when I say a long time – sharks do not just predate dinosaurs and crocodiles, but are even older than trees.† Teeth have been found from the late Ordovician period at least 440 million years ago, and there have been dermal scales found in sandstone 455 million years old that many argue are also from an early shark. Sharks have always been relatively common, have included some of the largest animals ever to swim our seas and survived through four of the five biggest extinction events in our planet's history.

When the first sharks began to appear in the fossil record, there were already some placoderm fish, such as *Dunkleosteus*, that had functioning jaws, but rather than teeth they sported a cutting edge to the bone of the jaw itself. There were cephalopods with beaks, other molluscs with rasping radula, crustaceans with claws … but the shark's hard, enamel-like tooth was a leap in the evolutionary arms race, the laser-guided smart bomb of its day.

Sharks lose and replace teeth constantly throughout their lives. A lemon shark discards 20,000 in its first 20 years of life, but may live beyond 50 years, and big great whites might get

---

* Tonic immobility can be induced in a variety of ways, including flipping sharks over onto their backs. It seems possible that orca might exploit this evolutionary Achilles heel, using it to immobilise sharks before eating them.
† The earliest trees appeared during the Devonian, 385 million years ago, and dinosaurs not until 230 million years ago. Sharks were around long before either.

through 40,000 in a lifetime, meaning the discarded teeth are plentiful. They are also rock hard and take centuries to erode – pretty much the model blueprint for fossilisation. Diving in well-used shark encounter hotspots – such as Fuvahmulah tiger shark harbour in the Maldives, or the quay at the Big Game Club in Bimini, Bahamas (where Ernest Hemingway used to machine gun the bull sharks from the dock) – you can swim along the bottom and sieve out teeth from handfuls of sand.

Fossilised teeth have also been found in riverbeds, in the Antarctic and in the very deepest oceans. These teeth are now believed to have evolved from the dermal denticles (the tiny teeth that cover the skin of the shark, creating micro-turbulence in the water and aiding their hydrodynamism). They develop in the throat and roll forward in waves, replacing the active layer at the front of the mouth as they dull and whenever they are lost. This can happen weekly. Even today, these teeth are some of the sharpest things in nature. I have discarded teeth in my 'dead things' collection at home that you could use for minor surgery and fossilised teeth that you could still cut yourself on even though they are millions of years old.

Proto-sharks such as *Cladoselache* had rounded teeth for snatching bottom-dwelling invertebrates, and more ancient forms still around today like the frilled shark have bristling, velcro-like needles.

It did take a while though for the highly specialised cutting tool you find on a modern great white to evolve. And when I say a while, I mean approximately a hundred times longer than primates have been around on the planet.

Frilled sharks still exist today in our deep oceans, while the sharks of shallow seas have changed immeasurably. This is probably due to the fact that the cool, dark, deep oceans are relatively constant environments, and great extinction-level events have affected them less than other habitats. Shallow seas, on the other hand, have been perhaps the most affected environments, particularly by climate crisis events. Radical changes and new environmental challenges drive rapid evolutionary change – organisms that fail to adapt fade away.

Even the hammerheads – which are generally considered the most recent shark group to appear – have been around for maybe 43 million years. The iconic cephalofoil of its head is believed to spread out its electro-sensing ampullae of Lorenzini, while providing more space between the eyes and nostrils, allowing for greater depth and distance perception both in vision and smell. You can also see some hammerhead sharks rolling or listing on one side, allowing the head to come into play, providing lift and greater energy efficiency in their swimming.

· · ·

Perhaps the closest thing you can get to diving alongside one of the Ordovician proto-sharks is to plunge into icy Alaskan waters on the periphery of the Arctic Circle. Here at night the shark relative the chimaera* comes up from the depths to feed. These ghost fish are not particularly big, no longer than my arm, but they are an odd throwback, with weird fins, extravagant pendulous noses and long-lobed tails. Unlike true sharks, whose upper jaws can protract forward to catch prey, chimaera have their upper jaws fused to their skulls. Many sport protective venomous spines. The oddest thing about them though is their eyes. Chimaera boast giant eyeballs for sucking in light at depth, and the most dramatic tapetum lucidum found in nature. Meaning 'bright tapestry', this is a layer of cells at the back of the eye that reflects light back through the retina, increasing night

---

* Though in the larger group that contains the sharks, the chimaera may have branched off as many as 400 million years ago.

vision. It is also what gives the 'eyeshine' when you shine a torch into the gaze of a nocturnal animal (emulated by the 'cat's eyes' keeping motorists safe on the roads at night). Diving in kelp forests at night, the light from your torch hits the chimaera that flit around you, shimmering bronze and silver from their flanks. However, when your torch hits their eyes, a dazzling laser beam of light reflects out into the water – they're the only animal I've ever seen do this. When Dr Evil of *Austin Powers* fame talked of 'sharks with frickin' laser beams attached to their heads', it was more accurate than he could ever have imagined …

Another denizen of the deep can also be encountered in the shallows at night in the nearby waters of British Columbia. Here, shy sixgill sharks can be seen, though only after many weeks of waiting, searching and baiting. We even built a custom-designed aquatic CCTV camera system in order to spy on them. We sat up day and night watching until we finally caught a glimpse of one at our bait and, panicking, dashed to get underwater. These slumberous sharks are prosaically named because unlike most of their kin, which have five gill slits, they have six. (There is also a sevengill shark. You'll never guess how many gill slits they have!) Sixgills are sluggish, cow-like brutes, with rheumy empty eyes and hacksaw dentition designed to get through the carcasses of whale falls and other bottom-borne carrion; they lock on with their sawtooth jaws, then swing the body around like a lumberjack trying to get through a testy tree trunk. These are in the most ancient of modern shark groups, dating back to the early Jurassic perhaps 200 million years ago.

The sharks of the Arctic are the most extreme of all fish, able to endure conditions that no other animal can. They slow their lives right down, living, maturing and growing at a pace that would make a snail yawn. The 'sleeper sharks' of the planet's coldest seas may be capable of bursts of speed (as has been seen by biologists that have tagged them underwater) but they rarely if ever use it, choosing instead to conserve their precious energy like subsea sloths.

One extreme exception is found in the velvet, chocolatey waters of Prince William Sound, an ethereal, mirror-like body of water reflecting snow-capped Alaskan peaks and bottle-brush Sitka spruce forests. Prince William Sound is one of Alaska's most famous wilderness areas, and also the site of its most notorious ecological catastrophe. On 23 March 1989, the oil tanker *Exxon Valdez* ran onto a shallow reef and discharged at least 11 million tonnes of crude oil into the pristine wilderness area of the sound. After three days of pitifully slow reactions, a storm blew in with 70mph winds, carrying the oil out over a vast area. Seals, sea lions, sea otters and seabirds were killed in their hundreds of thousands.

More than 30 years on, the sound has mostly recovered, and disasters like the *Deepwater Horizon* oil spill have made what happened here look like a drip of chip fat in a swimming pool. Now the sound is once again a haven for wildlife. Black and brown bears dig for clams on the foreshore, bald eagles gather in the treetops like so many crows, and orca and humpback whales course through the green waves. While just as stunning as southeast Alaska, where the cruise ships bimble round Juneau and Skagway, Prince William Sound is noticeably wilder. During one trip to Hinchinbrook, my companions and I never saw a single other boat all week. One afternoon our skipper had a few hours to himself and popped out for an hour. He came back with a halibut that weighed the same as me.

In the middle of the year, a different kind of fish attracts everything from apex predators to wader-clad human fishermen to the sound. All are summoned by the return of the salmon – chinooks as long as a 12-year-old child, pinks with their grotesque humped shoulders and hooked jaws, cohos or silvers that fight like caged demons, and stripy chum dogs that can battle 2,000 miles up the Yukon River as they return to their natal spawning grounds. These are red-flushed, punch-drunk brawlers. Their days are numbered, scales and digestive tract dwindling, sexual systems burgeoning. But despite the fact that they are already tatty and dying, they still face a furore of torrents and cataracts to reach their breeding sites. And in order to make it to the mouth of the stream they were born in, they have to run the gauntlet of assembled predators, one of which is an unlikely shark.

Ravencroft Lodge sits on a long peninsula stretching out into the sound, surrounded by old-growth forest, carpeted in dense green moss. It's a two-hour boat ride from the nearest town of Valdez, which is itself a half-hour flight out of Anchorage. Blue-black Steller's jays bounce between the log cabins, rufous-breasted hummingbirds buzz about feeders on the veranda, and every tree seems to resound with the zebra-like bray of the bald eagle. Everywhere you look you see mountains and marine life beyond compare. When the sun shines and the waters are calm, it is a realm of the gods, and the most tranquil paradise on the planet. It feels like you should whisper.

Ravencroft Lodge was originally set up as a retreat for hunters and fishermen. About a decade ago there was a brief craze for catching the little-known salmon sharks seen speeding in front of the lodge. With the kind of empathy that can only be gained by being out on one small patch of water for decades on end, the owner of the lodge, Boone Hodgin, inadvertently became a world expert in the behaviour of these 'mini-great whites'. However, within two years of modest catches, the sharks almost disappeared. Boone decided to find out more.

The salmon shark is a real biological oddity. Most shark species occur in warm or temperate waters. Comparatively few hunt the chilly seas this far north. Salmon sharks, though, use counter-current heat exchangers to run blood that's been heated by the movement of their muscles alongside blood heading to the brain, eyes and digestive tract. Hence their whole body is well above the ambient temperature of the water. To all intents and purposes, this is a 'warm-blooded' fish.

These heat-exchange devices are ubiquitous in extremophiles – the kinds of animals that exist in environments we humans would consider inhospitable. The scientific name *rete mirabile* translates as 'miraculous web', and they are used in the ears of elephants to shed heat and the nostrils of reindeer to warm freezing inhaled air. They are present in the flippers of leatherback turtles, enabling them to stay active in the waters around the British Isles, and in tuna, porbeagles and great whites, giving them the superpower of explosive speed in chilled waters that slow their prey. Many sharks eat what they come across; these predators can eat what they want. They have the ability to outrun and outgun their prey, and hence can select their targets. The blood vessels specifically warm the red muscles that run like pistons along salmon sharks' flanks down to their tails, powering the side-to-side movements of the tail and propelling them forward. This combines with a spindle-shaped body, crescent tail and conical snout to create a perfectly hydrodynamic animal that some sources suggest is even faster than the shortfin mako (usually considered the fastest shark on earth).

As no one had ever dived with a salmon shark before, Boone built a shark cage out of scrap metal so that he could further investigate the drop in numbers. It turned out to be unnecessary, as the sharks were totally oblivious to humans in the water. Diving with salmon sharks in Prince William Sound is not like any other shark dive. First up you don't chum the waters; salmon sharks are highly visual predators and don't seem to be as drawn in by scent as so many other species are. Instead you cruise about the mercury-calm waters of the sound until you see that V-shaped wave from a dorsal fin beneath the surface. When you find a shark swimming in circles, apparently warming itself ready to hunt, you get in and you hope.

These shark encounters are best had while freediving, keeping light and manoeuvrable, and within a metre or two of the surface. The physiology of the freediver has been studied now for half a century, and has yielded intriguing results which help us to understand our own bodies, our evolution and the potential of our underused abilities. Freediving is particularly in vogue now, with trained breathing being touted as a panacea for all manner of ills.

Humans possess certain advantages that aid us in our quest for the deep. Best understood of these is the mammalian dive reflex. Immersion in cold water results in us shunting blood from our extremities to our cores and dropping our heart rates; all functions that maximise the use of our onboard oxygen. You can gain some of these advantages merely by splashing cold water on your face.

On a recent diving expedition the entire team decided to give it a go. We sat round a table with a pulse oximeter on, and put our faces into a bucket of iced water. The results were immediate, with one of our team having their resting heart rate drop from 90 to 50 beats per minute in 30 seconds!

The idea is that on contact with cold water our blood vessels contract away from the surface of the skin (where heat would be conducted away 20 times more effectively than in air) and our nerves react to the immersion to reduce the contractions of the smooth cardiac muscles. This reduces our oxygen consumption, which is invaluable for any diving creature.

This reflex, and the facts that we have a relatively recent common ancestor with the whales and our skins are naked of hair like dolphins, combined with the preponderance of human settlements around coastlines, led to the proposal of the aquatic ape hypothesis in the 1960s, an elegant and popular theory that our ancestors lived semi-aquatic lives. After all, coastal seas are the richest in foraging fare, and proof that primitive peoples took advantage of this is found in mounds of munched seashells left in middens by our coastal antecedents. Evolutionary biologists have even posited that our upright bipedal posture might have evolved to allow us to wade in shallow seas to forage. Sadly, most of aquatic ape theory turns out to be entertaining bunkum, but the connection of our ancestors to the sea is very real, and perhaps nothing puts you more in the shoes (or fins) of the relic human than freediving.

With freediving, you leave the unwieldy scuba gear on the harbourside and merely step into the blue. It represents the most pure experience you can have of the underwater world. Liberated from the confines of scuba, you feel utterly free, able to bank, weave and roll. No safety stops, no bubbles, no kit to malfunction or physics to boil your brain with. Just you, your lungs and the sea.

Watching the masters of the art – like Guillaume Néry, William Trubridge or Tanya Streeter – interact with their aquatic environments underwater makes it seem completely natural, like we were meant to be this way. They are like marine mammals in skintight silver suits. And being free of the tank and blown-out bubbles completely transforms how animals relate to you underwater, allowing these top freedivers to dance below the waves with their marine mammal cousins in ways too beautiful to put into words. To freedive alongside a super-powered salmon shark beneath snow-covered peaks would surely be the experience of a lifetime.

We had arrived at Ravencroft off the back of a few weeks of solid rain, which had left the sound green with glacial silt and plankton blooms. There were thin slivers of cloud spiralling between the trees, twisted by an imperceptible wind. Four sea otters with golden grizzled heads spyhopped out of the water with their paws held up in surrender, watching us inquisitively, and to our right a pod of monochrome Dall's porpoise sped over the surface, their fins scything an explosion of backlit water droplets. Otherwise the sound was so calm it was otherworldly.

We found ourselves flying over fields of gigantic white plumose anemones, full of friendly wolf eels and giant Pacific octopus. And then in front of our skiff a V-shaped wake appeared. There was something ominous about it, a sense that below the surface there was something big and powerful, moving languidly and easily, and holding back its explosive destructive potential – a Bugatti Veyron rolling in to a traffic light, engine growling. And then the black dorsal fin broke the surface, and a thrill shivered up my spine.

We pulled on our wetsuits and fins, pulses racing, and slipped quietly over the side. Underwater, the salmon shark resembles the great white that many people see in their nightmares: coal and graphite above and blotchy white below, huge empty black eyes and rapier teeth that spill from their mouths. Most that we saw were small; less than two metres in length. We did, however, see a couple of big females that were more like three metres, bulky and intimidating. But they couldn't have been less interested in us. Though there are isolated reports of them taking sea otters, this shark is a fish-feeding specialist. In fact a careless movement will spook them, and they'll be off with a lightning-bolt swipe of their keeled tail.

Everything about this unique encounter sums up what is so special about sharks. Physiologically adapted to such a broad array of habitats and challenges, they are supreme hunters that can thrive anywhere from glassy tropical waters to gloomy frozen seas.

. . .

Although there is a very definite 'type' that most people think of when they hear the word 'shark', sharks actually exhibit a remarkable range of shapes and sizes, as well as unusual physical attributes that set species apart. One of the most memorable of all my shark encounters exemplified this, and was also one of the hardest won. The thresher shark is a timid creature that has to be approached with a lot of care and attention. At present, the only known way to see them is to dive very early in the morning, when they come up from depths to visit cleaning stations.

My crews and I have attempted thresher dives in several sites in the Maldives and Philippines, with varying success. But when it went right, off the Philippine island of Malapascua, it was memorable indeed. From the dawn gloom deep-blue, an alien form glinted and glimmered. I hung motionless above the seabed, barely even daring to breathe in case my expelled bubbles spooked the ethereal shape just beyond my gaze. Then, languidly, lazily, the metallic torpedo turned towards me and started to focus into view. The shark was perhaps four metres long. Half of that was made up by a scimitar-shaped tail that trailed behind it like a silver banner in the breeze. The large, light-gathering eyes were billiard-ball black; the whole form of the fish seemed cloaked in aluminium foil. As its mirror flanks sinuously twisted side to side, it caught the early morning light, and suddenly the thresher shark was revealed in all its bizarre, brilliant glory. It was one of the most overwhelming wildlife encounters I've ever had, with a shark we have right here in British waters.

The most evident feature of a thresher is its tail, which can be more than half the length of its body. Aristotle, in his *Historia Animalium* in 350 BCE, was the first to suggest that this mighty tail must be used as some sort of whip to lash out at and stun prey.* Every biologist and diver since had taken it for granted that this is what they did, but no one had ever actually seen it happen. It wasn't until July 2010 that German filmmaker Klemens Gann became the first person to film it.

'It was on Pescador Island, a small island maybe two kilometres offshore from Cebu, the Philippines,' Klemens told me. 'There is a shallow reef plateau and a wall that drops down to 50 metres.

'I was super lucky to be the only diver in the water that day. Threshers are super shy, and move away if they sense you are there. This one, though, was focused on a shoal of sardines, and didn't seem to see me.

'I'd watched them hunting before, and knew that once they are heading in to the fish, they don't care, so I wait, and wait, then as it heads to the fish, then and only then do I swim in.'

Watching Klemens's footage back, you have to hold your breath, much as he must have done. The shark comes in slow and cautious, then at the last second it seems as if the footage has been speeded up; four or five rapid sweeps of the tail and it is under the giant silvery baitball of sardines, and then whack! It jackknifes its whole body, and the tail slices up through the water into the fish. A few fish are hit, discombobulated, and the thresher lazily turns back to suck them down.

It was the slicing-up bit that took me by surprise; I always assumed the thresher would slap its tail from side to side, as it does when it is swimming (and indeed later footage showed

---

* He also gave them their scientific name of *vulpinus* – the fox, as they are considered cunning and clever.

a thresher slapping a bait box in just this manner); however, it actually makes far more sense to hunt this way – using the more streamlined leading edge of the tail to cut through the water, therefore incurring less drag and maintaining its momentum. The wonder of evolution and deep time is present in the physical form of the thresher, but in the moment any diver is just awed to be in the presence of such an improbable masterpiece.

●　●　●

While the tide is turning when it comes to the rhetoric around sharks, this is still an animal group that remains much maligned, and even more practically persecuted. A perfect example is the blue shark. I've been lucky enough to spend time with them in three different oceans. They are to my mind the most interactive and sentient of all fish. The blue name comes from the ethereal, almost otherworldly blue colour of their flanks, seemingly lit from within. Their long, elegant forms move with hypnotic waves, making them one of the few shark species you could accurately describe as objectively beautiful.

In 2013, a male shark off the coast of San Diego spent the best part of two hours with us, and rather than focusing on the food in the water around him, he constantly pressed his snout into our hands, seeming to enjoy the sensation of having his highly sensitive ampullae stroked.

Off the shores of the UK, blue shark dives, while unpredictable and challenging, are to my mind the UK's greatest wildlife encounter. In 2022, I had one day with perhaps 30 sharks flitting around us like over-inquisitive poodles. Their personalities were palpable, some nervous and never coming to within metres of us, others bolshy and bumping their cohorts out of the way or bossing food. A few sought the sensation of physical contact.

One female bore conspicuous mating scars (females' skin can be five times thicker than the males in order to ward off these mating bites) and was as pugnacious as any individual shark I've ever seen. She had to be pushed off multiple times to avoid one of us getting a nip.* Each animal had a different character. By the end of the day, the thought of losing even one of them seemed like an impossible tragedy. Yet one fishing boat could have come through and taken the lot in one go.

Blues take a long time to grow and to mature. At 11 months they have a longer pregnancy than humans, and they produce a small number of pups. They are the epitome of a species that is incredibly vulnerable to overfishing. And yet overfished they definitely are. This is the most fished species of shark worldwide (with perhaps 20 million individuals taken each year) and the most common found in Asian fish markets. In the Med, blues are listed by the IUCN as critically endangered.†

The blue shark is a wide-ranging species that can be found in temperate and tropical waters, and is usually caught as bycatch in the tuna and swordfish fisheries. Most are finned and discarded. Official figures show the global catch as being below the presently set catch limits. However, independent research in Hong Kong fish markets has shown that blue sharks are three to four times more prevalent than they should be. This suggests – as is common in fisheries' data – that catch sizes are being widely and wildly underestimated. Despite all this, the Marine Conservation Society (MCS) states that 'current assessments indicate no concern for fishing pressure'.

---

* In the summer of 2022 off the coast of Cornwall, one of these dives was the subject of arguably the UK's first shark bite. Media hysteria obviously followed, though the swimmer was desperate to point out that the bite was minor, and that the sharks posed no threat (and a huge thank you to her for that).
† Near threatened worldwide.

This pressure spills over to affect the most glamorous of shark relatives, the ray. Essentially the 650-odd ray species are flat sharks that look a bit as if you've taken a typical shark and run it over with a steamroller. They share with sharks the cartilage skeleton that is lighter and more flexible than bone, the internal structure, and the large oil-filled liver to compensate for the lack of air-filled swim bladder sported by bony fish. Behind the eye, rays (and some bottom-dwelling sharks) have an opening called a spiracle which can be used to pump water and hence oxygen into the body and over the gills. This enables them to breathe while lying motionless on the bottom, without sucking in sand and other detritus. The fact that species which swim up in the water column still possess these spiracles is just one of the reasons that we believe the rays as a group adapted to a life on the bottom. Mantas and other mobulids would then have taken a different course, and returned to a life of constant swimming.

The two species of manta rays can be found throughout the world's tropical seas, with oceanic mantas also venturing into cooler waters at higher latitudes. Over time, regional populations have tended to become isolated and have evolved with subtly different traits. Certain populations target particular resources that change throughout the year, and oceanic mantas range over vast distances out into the deep blue to find hotspots of food.

Mantas are often said to have the biggest brains and encephalisation quotient* of any fish. Off the coast of Ecuador, I had the memorable experience of having a vast oceanic manta (five metres from wing tip to wing tip) swim over my head, blotting out the sun. Then, as he was directly above me, he stopped swimming, and sank down onto my head. The bubbles from my scuba regulator poured out of my mouth and surged up his belly

* Encephalisation quotient, or EQ, describes intelligence in animals by measuring the proportionate size of body to brain.

and over the sides of his fins. The giant animal remained there motionless for perhaps five minutes, clearly revelling in the novel sensation of a submerged Jacuzzi bubble bath.

I had an encounter of equal drama but radically different scale sailing from the Mexican port town of La Paz, near Cabo Pulmo. Not long after dawn, the serenity of the sunrise was broken by the thunderclaps of breaching mobula rays alongside our boat, leaping one after another in synchronised displays.

These mini mantas aggregate here in their thousands as they head north along the coast to their breeding grounds, bedazzling visitors with their aerial acrobatics. No one is 100 per cent sure why they breach. There is certainly a pragmatic function, as clattering back into the water helps blast parasites from their skin. It is also an effective anti-predator strategy, deterring the orca, bottlenose dolphins and tiger sharks that would make them lunch. However, it seems likely to have a social function too. The sound of the slaps as they land back in the water carries for hundreds of metres – could they be using them to communicate, as humpbacks do? Are they training?* Or is it a display of fitness – a classic piece of sexual selection, as animals show off their prowess and hence genetic superiority?

To answer the question, we headed over to the centre of the breaching mass in our inflatable boat and filmed the leaping mobula in super slow motion. Like all sharks and rays, these mini mantas have their sexual organs on the outsides of their bodies, and it was clear it was both male and female mobulas breaching. This made it seem an unlikely display of fitness. Usually this would be performed by just one sex – almost always the males – showing off to their would-be mates.

---

* Young cetaceans such as humpbacks will breach over and over again to build up their muscles, effectively working out to prepare themselves for their epic migrations. Potentially this breaching in mobula could serve the same function.

On expedition heading south out of the Sea of Cortez, we decided to put some time into trying to locate the migrating mobula that were due to be arriving in the area. To try to find them using an eagle eye in the sky, we connected with a local pilot named Siddharta. Our excitement turned to bemusement when he landed alongside us.

'What the hell is that?' I commented, as his curious-looking micro-light circled above us. 'It looks like he made it in his garage.' As it skidded in to land on the water, we could see it was a fibreglass hull attached to some scaffolding poles, with what looked like a big barn door as the wings and a couple of bathtubs for its floats. The prop was situated centrally behind the wings and appeared to have been salvaged from a plane crash.

'Amazing plane,' I complimented him, as we got close in our dinghy.

'Thanks,' he replied, 'I made it in my garage.'

Sid told us that the giant aggregations of rays had not yet made it as far north as La Paz, and that we would have to head south nearly 100 miles to find them. He then told us that he had flooded the engine and needed help to get airborne again. 'I just need you to stand on the float,' he told me, 'just to weigh down the front of the plane, and when the engine gets going and she starts taxiing, you need to jump off into the water.' I stared at him with my jaw hanging open as he hung off the prop and tried desperately to get it to start. Sadly it was thoroughly flooded, so we unceremoniously towed him in to the coast using our dinghy, all the billionaire owners of the posh yachts looking on in disbelief at the junkyard float plane.

When we finally did see them from the air, it was with our drone not a homemade plane. From above, the swarms of mobula look like windblown autumn leaves, swept around in billowing clouds the size of football pitches. Just like with their bigger manta cousins, the finest encounters with mobulas can be triggered artificially. They are not attracted to bait, as a shark would be, but can be drawn in by light. The first time I tried this was 25 years ago, off Hawaii. Locals had just discovered that if you dropped light to the seabed, it would offset an artificial plankton bloom, and within minutes the mantas would start to arrive. For those early experimental dives, they butchered the lights from a motorbike, plugged them into a truck battery, and took the whole lethal assembly down wrapped in plastic and gaffer tape.

My first of these dives was a life-changing experience. Perhaps 50 mantas, each like Aladdin's flying carpet, glided over, around and at times straight into me, tossing me along the bottom like a windblown crisp packet. The biggest individual was a huge female they called Big Bertha. Even though she was identified as a reef manta – the smaller species – she was estimated to be five metres across.

Fifteen years later I returned to the same spot. Much had changed. Now I dived alongside half a dozen other parties, all of whom had come out to witness this magnificent assembly of spaceship-like creatures. Instead of poorly waterproofed headlights, they just took down a crate of dive torches. And Big Bertha was still coming in to feed. It seems she is still seen there even now.

The last time I experienced the manta vortex by night was in 2021 in the Maldives. It was an experience that played with my senses. With torch beams cutting through the gloom, and the balletic, almost computer-generated motion of the mantas around me, it was only an ambient soundtrack and our breathed-out bubbles away from being an Ibiza super club.

The Maldives is one of the few shark sanctuaries in the world, where the catching or landing of sharks and their relatives is strictly illegal. However, no more than a few hundred miles across the ocean in Sri Lanka is the world's largest manta ray fishery. These intelligent giants are caught in their thousands, both for their meat and for the spongy material of their gill rakers, which are used in traditional Chinese medicines for things like asthma (for which there is no evidence to suggest it might work). While many NGOs battle to bring this trade to an end, their struggles have not been aided by the WHO legitimising traditional medicines and affording them an equal accepted status alongside means-tested modern medicines.

The concept of the shark sanctuary is one of the great potential wins in elasmobranch* conservation. The world's first was Palau in Micronesia, set up in 2009, with no-take zones designed to allow shark numbers to rebound in the area. It helped that Palau gets much – even most – of its economic turnover from dive tourism. The idea of the shark-diving tourist began to gather momentum, and soon 16 other nations around the world followed suit.

I did my first shark conservation programme while working for National Geographic in the year 2000. We worked with legendary shark biologist Dr Sam Gruber at his ramshackle institute the Shark Lab, a bunch of bright blue weatherboard shacks to the south of the elasmobranch hotspot of Bimini, the smallest habitable Bahamas island. It was the paradise location for Ernest Hemingway's *Islands in the Stream*, after he spent his summers here in the 1930s, fishing for huge game fish and swilling rum. It's only 50 miles from the Florida mainland, but was a startlingly quiet piece of paradise.

Sam's long-running projects were instrumental in the whole nation of the Bahamas becoming a shark sanctuary in 2011. Shark tourism grew so rapidly thereafter that studies soon showed an

---

* The broader fish subclass that includes the sharks, skates and rays.

enormous benefit to the economy. Approximately $100 million a year now comes in directly from shark tourism, with some studies showing that a big and long-lived shark may be worth more than $1.9 million throughout its lifetime. That same shark might garner only $500 if caught and sold for meat.

Protecting somewhere like Bimini in the Bahamas can have a disproportionate effect on a shark's chances of survival. Knowing that they rely on the mangroves, reefs and seagrass meadows for essential parts of their life cycle, this comparatively small haven is critical to the survival of perhaps 20 species.

The most important of these is surely the critically endangered great hammerhead. Bimini is the best place in the world to see them, and the easiest. In the early months of the year, you can see them a short swim 200 metres offshore in South Bimini. In my decades of diving here I've learned to identify individuals by their markings, size and other distinguishing features. In particular, there is one giant female called Gaia who has been returning most years for nearly a decade. She is bold but gentle, interactive, stunningly beautiful and, whenever encountered here in these early months of the year, always appears unusually deep-bellied. In 2023 scientists managed to ultrasound her as she swam by, and revealed she was indeed heavily pregnant, carrying around 28 pups.

Gaia has been my lucky charm, and the wild animal I feel I know and love best. However, like most sharks that come here, Gaia is a nomad and must eventually head away from these protected seas, heading northwards up the coast of America, where it's believed she and the other mothers deliver their pups somewhere off the Carolinas. Once out of the sanctuary, this magnificent and critically endangered animal that I have learned to treasure like a friend can be legally caught on a rod and line. She could be served as swordfish steaks or chucked back in to sink unmourned to the seabed. Every season, when Gaia or any of her fellows are first seen, the shark divers of Bimini crack open a few cold beers in celebration.

Sanctuaries work. But this also makes them a target. For years now, hundreds of fishing vessels have plundered the Galapagos sanctuary, with 73,000 hours of illegal fishing within the supposed haven every single month. On my aforementioned trip to the Maldives at the end of the pandemic lockdown, we visited 30 Indonesian vessels in Male harbour. The boats had been impounded after having been caught illegally fishing within the sanctuary. It was difficult to contain my fury with the fishermen until I chatted to them. They were trapped on board their small boats, bored, starving and frightened. They'd been there for over a year. None earned more than 50 dollars a month. Their job was to undertake illicit raids into banned waters, returning with their bounty to Chinese factory ships that waited just outside the borders to receive their catch without any risk to them. Unless these unscrupulous factory ships and the big businesses that utilise their catch can be brought to task, the arrest of foot soldiers will have no effect.

That protection also has to extend not just to fisheries but to marine environments. When I first visited Bimini, Dr Gruber took me on a small skiff into the seagrass meadows and mangroves of Mosquito Point. Dr Martin Luther King, looking for a place of tranquility, came here to write his Nobel Prize acceptance speech (there is a bronze bust of him now deep in the heart of Smuggler's Pass), and it's easy to see why. The mangroves still the waves and wind. Herons and egrets stalk the root systems, bonefish snort along the sands, and terns dive for baitfish.

Mangroves are a biological anomaly, able to shed salt through their leaves and survive in low-oxygen environments no other plant could stand. Red mangroves even have the ability to filter and desalinate water through their roots. The mangroves are a vital habitat for our oceans, dissipating the energy of increasing storms, depositing nutrient-rich sediments, and capturing atmospheric and oceanic carbon. Their heavy roots also dig down into the sediment and bind it, breaking up water flow, and the labyrinths of their root systems create nooks and crannies which are the nurseries for so many of the most economically important and ecologically necessary animals of our reefs and open oceans. It is therefore extraordinary that we have taken so long to wake up to the importance of mangroves and to place real economic value on them.

Dr Gruber had a long-running project showing that these mangroves were an essential nursery for certain species of sharks, including the lemon sharks we would later dive with further out to sea. When I returned in 2005, Gruber's research had revealed that those lemon shark pups would live there in the mangroves for perhaps the first four years of their lives, safe from the attentions of predators (including adult lemon sharks). We returned again in 2007 to the news that mature female lemons were bound to the mangroves where they were born, and that however far they roamed they would come back to the same spot to have their babies. However, in 2011 the story had soured. Much of Mosquito Point had been taken over by a consortium of Asian companies, who were dredging the mangroves, coral reefs and seagrass meadows to make way for a vast hotel, casino and luxury home complex.

Now Mosquito Point is no more, but the dredging continues. A huge pier has been ploughed through the coral reefs to accommodate gargantuan cruise ships, which sit smoking in the dock and turning Bimini's blue skies grey. As an outsider it's difficult to get involved. There are Bahamian people who work in the complex and support the construction. Certainly local

councillors have been vocal in supporting it, slamming conservationists who mourn the passing of such natural jewels. My
Bimini friends, though, hate Resorts World and what it has done
to their island. 'I used to come here as a kid,' says a friend, 'to
freedive for conch and catch bonefish. But now I'm not even
allowed here. We've killed off the thing that made people want
to come here in the first place.'

Wanting to learn a little more about how increasing extreme
weather events affect this low-lying island group right in hurricane highway, I asked my friend what it was like to be here in a
big storm.

'I was here in Hurricane Andrew,' he said. 'It really wasn't all
that bad, we weren't hit that hard.' That was unexpected.

'The storm hit the mangrove and just kind of broke up,' he
went on. 'It'd be a different story now.'

And for the sharks too. In what is supposed to be a sanctuary,
profit-seeking developers have destroyed the place that the sharks
need to breed – when those lemons return here to give birth,
their safe haven will be gone.

• • •

My relationship with sharks and the media is constantly evolving.
In preparation for that first programme back in 2000, I spent a
few weeks* researching shark attack statistics and was staggered
at what I found. According to the University of Florida's Shark
Attack File, which was the only resource recording attacks at that
time, the previous year only three people worldwide had been
killed by a shark. Over the previous ten years there had not been
a year with more than ten fatal attacks. I had just come from
Malaysia, where they had talked about three people being killed

---

* Back then research meant actually going to libraries and reading books, or
calling scientists up on the (gasp) phone!

by falling coconuts each year! I put that into my piece to camera. A few years later there was a report saying that ten people had been killed that year by falling vending machines, so I changed my reference to that. In 2014, it was reported that 12 people had been killed taking selfies that year, so that became my go-to. You're hundreds of times more likely to be killed by a bolt of lightning, and champagne corks (12), dryer lint (5) and electric blankets (30) are all more fatal to humans than sharks!

The number of shark attacks per year might fluctuate a little, but the fact is that sharks just do not have a major impact on human mortality anywhere around the world. Even in places where they have grabbed the headlines, such as Réunion or New Caledonia, the shark is not a significant cause of human deaths.

So why do we fear them so much? Well, some of it is likely related to our primal fear of deep water, an environment where many of us feel vulnerable and where monsters, it seems, may lurk beyond our gaze. However, much of this fear is surely down to the media, and one film has a lot to answer for. (I'm sure you can already hear those two ominous chords in your head!) Like every child of my generation, I saw *Jaws* when I was way too young. And when I say way too young, I mean I was so small I could barely pick up the cushion I buried my head in.

'Jaws' is often considered to be a dirty word now by conservationists, with the iconic film and the many abominable sequels being credited with all public hatred of sharks. And they have a point. Peter Benchley, who wrote the novel on which the film was based, famously became a vocal shark fan and said he wished he'd never written his novel. Steven Spielberg, who made the film, has recently joined him in recognising the negative impact the film has had.*

---

* It remains in my top ten films of all time, with the drinking scene in the *Orca* far and away my favourite piece of scripting and acting in Hollywood history.

Sharks had little place in the broader public consciousness before this 1970s classic, but after it they were perceived as malevolent monsters that would persecute and pursue their human foes. Just to be in water where sharks might be was to risk being instantly and horrifically eaten. This dread we have of sharks is part of the reason why it has been so hard to get the public to wake up to the real shark crisis: the fact that we humans are taking at least a quarter of a billion sharks from the world's oceans every year – a number that cannot possibly be sustained.

Despite many people now having a more enlightened perspective on sharks, their position in the media remains that of demon incarnate. A shark attack in Australia is front page news in the UK, where a fatal car crash might not make the news at all. Part of that is because a shark attack is so unusual. It also plays into our fascination with the macabre, the grisly, the bloody. (Any alien who spent time flicking through Netflix would be convinced that the main cause of human death was serial killers.)

Some of this press is completely understandable – the story of a swimmer bitten in half by a great white shark is utterly horrifying and plays into our primal fears of nature, predators and deep dark water. Much more of it, though, is just purely self-serving sensationalism on the part of bad journalists. Every single year the front pages of the red tops will light up with tales of swimmers or paddleboarders narrowly escaping death at the teeth of a demon shark. All too often, the culprit is in fact a toothless basking shark. As I write this I have in front of me a *Daily Mail* headline about a British tourist suffering a terrifying attack by a great white 'mere feet away from him'. You have to read down a paragraph to find out the shark was mere feet away on the other side of a shark cage. Memorably, in 2021, the front page of the same paper showed a picture of the first great white shark in British waters, surely about to swallow a bunch of swimmers.

Closer investigation showed the photo actually showed not a fin, but a sewage pipe sticking up out of the water.

If the most powerful dogma in the public perception is that an animal is a maneater, vicious, bloodthirsty and demonic, it follows that many people would be scared of it and want rid of it. The idea of conserving sharks would seem laughable to many people – the equivalent of protecting leeches, mosquitos or pubic lice.

I have been making television programmes about sharks and the issues they face for more than 25 years. We've covered every single angle of the complex shark extinction problem, from the use of squalene in new vaccines and lipsticks* to the research conducted by the University of Exeter showing that deep-fried shark is on sale in British chip shops as flake, huss, bull huss, rock or rock salmon. One of these programmes was on primetime television, when I dived live with hammerheads, bulls and reef sharks in the *Strictly Come Dancing* time slot. We talked about how mangrove destruction harms shark nurseries, and how long-line fisheries decimate non-target pelagic sharks. However, it was clear that most of our audience didn't hear any of that stuff. In fact, most of the tens of thousands of responses on social media just said things like 'Wow, I didn't know sharks could be beautiful!' and 'How is he not getting eaten?'

A lot of my social circle are divers, many of whom have already swum with sharks or would like to. While this is a growing subsection of society, most people in the world are not aware of the problems sharks face. Making people around the world love sharks is still the great challenge.

My cameraman friend Simon Enderby dropped in behind a finning boat passing through Sipadan Island in Borneo (a site so special they are seeking UNESCO World Heritage status).

---

* Shark liver oil, or squalene, is found in a huge array of products, and is rarely clearly labelled as such, including in flu vaccines.

He filmed the horrifying spectacle of sharks lying over the coral heads, their fins and tails freshly sliced off. The animals were still alive, slowly asphyxiating, surely in agonising pain. I, meanwhile, have stood in Asian and Spanish fish markets surrounded by thousands of bloody shark carcasses, the sum of just a single day's worth of trade. The floor runs red with their blood like a scene from *The Shining*. In Mexico and Indonesia I've walked past kilometres of racks of drying shark fins, and seen the windows of shop after shop in Hong Kong and Shanghai filled with those same fins, bagged and ready for sale. I've seen mighty whale sharks robbed of their dignity as they bleed out on a dock and odd species I don't even recognise laid like sardines on the slab in Latin American markets. I've come across great white and manta meat openly on sale by street vendors within protected sanctuaries, and tigers and hammerheads grotesquely hoisted on a rack. Even in UK ports and fish markets I've witnessed crates and crates of baby sharks a decade short of maturity, plus blues and makos bound for fish and chip shops or set aside as 'rubbish catch' for use as fertiliser or fish food.

It's clear that sharks are in big trouble, and even now a large part of the solution has to be addressing public perception of their importance. Thankfully, that's something my (often toothless) industry can do something to address, giving me some hope that not all of our work in the field will be in vain.

CHAPTER SEVEN

Great White Hope

If you're going to dive with a great white,
you have to own the water.

José Antonio Aguilar

*Out of the deep, deep blue, the most iconic, most chilling silhouette
in nature starts to coalesce: the great white shark, a giant at five
metres in length. Pectoral fins drop slightly, the tail moves in slow
motion, the white underbelly that gives it its name almost seems to
glow. The animal is broad, deep-bellied, criss-crossed by a railroad
of cicatrix, white on the dark upper side, pink on its belly. This one
is a female, and some of these scars are from the mating attentions
of male sharks, biting at her gills to try to pacify her, the shark
equivalent of a passionate embrace. Other scars are from the last
desperate throes of dying prey. I make eye contact with Simon, my
cameraman – a great bear of a man, who I only hope will make a*

143

bigger and more appetising target than me. Simon nods and speaks into his microphone.

'This is the one, Steve, go for it.'

Rolling the bars aside, I shuffle forward to the edge of the cage, my legs and fins dangling out over the abyss. I let go of the bar that was my last sanctuary, and swim out to join her.

• • •

It is no mystery why the great white shark is an object of fascination and horror for so many people around the world. In appearance and proportions it is the most frightening predator on our planet.* The reputed world record for a caught and measured white shark was off the coast of Cuba in 1945. It was 6.4 metres in length, more than three tonnes in weight, and sepia photographs show it with what appears to be a whole classroom of school kids sat on its vast corpse.† A great white's mouth bristles with 300 teeth, and in breaching attack it can slice a seal clean in two.

As with other shark species, most of what we know about the modern great white shark's (*Carcharodon carcharias*) evolution comes from studying those famous teeth and the teeth of their ancestors. It's a necessity, as the rest of their bodies are not prone to fossilise, and the teeth are found in such abundance. However, this leads to inevitable gaps in our understanding. You can't make accurate estimations of size, fin placement, sex or much else other than prey type just by studying teeth without an attached body. Throughout my years of studying biology, our understanding of megalodon, for example, has varied enor-

---

* Though I would argue that it is not actually our ocean's top predator.
† There are countless accounts of monster sharks up to 11 metres in length, but none have been scientifically measured, and most of these records have been actively discredited.

mously. Once considered to be part of the *Carcharodon* group and an ancestor to the modern white shark, they are now placed in the extinct family Otodontidae and not considered to even have been a close relative. Back in the 1980s they were believed to have been the size of blue whales. This was then radically revised by scoffing biologists to a fraction of that bulk. Now, though, paleontologists have swung back to a maximum size of about 20 metres in length and 110 tonnes in weight.

Another complication in figuring out much about either the meg or the ancestors of the white shark using dentition is that modern great whites' teeth change throughout their lifetimes. Juvenile white sharks have thin needle-like teeth (very like those of a mako), with lateral cusplets (mini teeth alongside a main tooth) that are specialised for catching fish prey. However, once the shark gets to three to four metres in overall length, the cusplets disappear, and the tooth broadens into a serrated cutting tool. This enables them to switch to feeding on marine mammal prey (though fish prey is still vitally important, and some populations or individuals might not switch at all).

Just as no mountaineer begins with a summit of K2, no diver starts with great whites without the safety of a cage. If you have any sense, you start with more manageable species in predictable situations and work your way up to it. I had the best part of two decades diving with humble reef sharks on tropical coral reefs before trying something bigger (bull sharks in Belize's famous Blue Hole), and then another decade before thinking it might be possible to swim alongside a great white shark. And even then the first step was to learn about great whites from behind the bars of a shark cage.

It was 1999, while filming for my first conservation series for the National Geographic, when I travelled to Gansbaai in South Africa, the home of great white diving, to find out how shark tourism could be a part of the solution to the problem of shark persecution (as it has proved to be in other parts of the world).

Shark fishing started in Gansbaai in earnest in the 1930s, and by 1940 was up to 1,000 tonnes a year, much of which supplied the insatiable demand for shark liver oil, used as a source of vitamin A and also as a lubricant. The fish factory provided a brief boom in the economical fortunes of Gansbaai, but then the small town faded back into obscurity until cage diving took off in the 1990s.

Back then, rather than this being a part of the canon of global wildlife encounters, it felt like a macho thrill-seeker's enterprise. People would run with the bulls at Pamplona, cliff dive in Acapulco and then come here to cage dive with great whites. Home-welded cages hung off the back of the boats, and shark wranglers would pull the great whites up to the stern, get their heads out of the water and pull their snouts back to show off their teeth. Rapidly the business expanded until there were few people in Gansbaai who didn't make a living directly or indirectly from the great whites. It wasn't particularly seasonal, either, with dives on any day of the year, weather conditions allowing.

The sharks come here to feed on the 60,000 Cape fur seals that pull ashore on Dyer Island. The strait between there and Geyser Rock is known as Shark Alley. On my first few filming trips to the Western Cape, we dived in murky waters, with the creepy forms of the sharks lurking beyond our gazes before they suddenly and eerily hoved into view, their big black eyes seemingly looking right through us. On one rare occasion, we had glorious 20-metre visibility and 12 great whites circling about us throughout the day. That experience was most memorable, though, for a fur seal pup swimming into the cage alongside me. It had a minor wound to its tail, clearly having narrowly escaped predation from a shark, and was utterly terrified. We eventually reversed the boat up to the kelp forests and shoved the little one out into the safety of the seaweed.

Cage diving has never sat entirely well with me for many reasons. Firstly, from an aesthetic perspective, I'm not wild about seeing animals through bars. Secondly, although I've been lucky enough to work with true legends of the shark game – people who really care about the animals – there are an awful lot of cowboys in the cage diving trade. They're in it for the bucks alone, and 'wrangle' the sharks mercilessly for photos and glory. This has led to the third problem, which is that there have been some high-profile incidents in recent years where over-zealous shark wranglers, hoping to drag sharks nearer to the cages so their customers can get the best photos, have actually led sharks into cages through their slot-shaped windows. The terrifying footage of thrashing sharks inside a cage where cowering divers battle for their survival has done the cause of shark conservation no favours. What has been much less publicised is that in every such case the shark has been grievously wounded, and in several cases has been documented as having died of its injuries.

Even without getting wet, great whites hunting Cape fur seals around their colonies is one of the great predatory spectacles in nature. During the daylight hours seals gaze with their great dark eyes down into the depths and have the edge over the sharks. However, for an hour or so at dusk and dawn, the seals' silhouettes are clearly visible from below, and the sun is low enough in the sky that the seals can't spot the sharks. This is when the hunting takes place.

Scientists in Australia wanting to get to the bottom of why the sharks are so choosy about the time of day that they hunt tested 44 different animals approaching bait at dusk and dawn. They found that the predators approached from the east at dawn and the west at dusk. The sun was always behind them, unable to penetrate far into the murky depths at those times. The reason for this could be to avoid overstimulating their own hyper-sensitive retinas, or to better illuminate their prey. Perhaps the greatest driver, though, is to enhance their own camouflage.

It's often a young seal, cavorting at the surface, heading out from the rocks to sea or returning to the safety of the shallow waters, that falls victim to this strategy. Perhaps it thrashes a little bit more than the better coordinated adults, attracting the attention of a shark's highly honed lateral line.* Whatever, you'll suddenly catch in your peripheral vision an explosion of water and the seal flying vertically up in the air, smashed upwards by the thrashing teeth of a monster great white.

Over the years I've filmed countless great whites breaching, the largest of which was more than five metres in length. An accelerometer placed within a decoy showed that seals are subjected to around 5G at the point of impact, the equivalent of a fighter jet in a tight turn. The conditions for breaching are so precise, and the window for it so short, that some biologists have an almost preternatural understanding of the process. When filming off Mosselbaai, biologist Enrico Gennari was standing at the rear of the boat watching the decoy being dragged behind us. He glanced up at the sky, looked at our position, then loudly started to count. 'Three, two, one,' he intoned. On the 'one' the water erupted behind us, and the biggest breach we'd yet seen happened. It was as if Enrico had an extrasensory connection to the beast waiting for him beneath the waves.

While great whites are generally helpless before the persecution of human fisheries, there is some evidence that those

---

* A channel that runs the length of many fish's bodies, which is filled with tiny hairs that can detect movement in the water.

hunting around the seal colonies here are learning to adapt to the human world. A bustling town has sprung up around Mosselbaai, casting its artificial light down onto the waters around the seal colony. That perfect moment when the sharks could remain camouflaged and the seals unable to perceive them was extended by as much as an hour a day. We continued seeing sharks breaching well after the sun had gone down. They'd been given an extra gift by humans – one of the few times people have ever done anything to help sharks.

In recent years, though, we have not returned to South Africa to film sharks, for the simple reason that there are no great whites there to see. In January 2016, two male orca were seen swimming south from False Bay towards Danger Point. Though they were males, their giant dorsal fins were not evident as they flopped to the side, one to the left, the other to the right, causing locals to nickname them Port and Starboard. Soon after, shark sightings up and down the South African coast simply ended.

In April 2016, marine biologist Alison Koch was called to investigate the carcass of a sevengill shark which bore tooth marks on its fins, and had had its liver neatly excised.

Over the next few weeks more sevengill carcasses turned up. The great white sightings decreased, and in some months dried up completely. It was not until the following year though that Port and Starboard were actually seen at Shark Alley. Just days later, the first dead great white washed up nearby. There were then no great white sightings from any of the eight shark dive operators for a fortnight.

The orca were spotted several times in late April 2017, and then a five-metre-long mature female great white washed up dead, its liver neatly sliced out and presumably consumed. Soon a good-sized male with the exact same wounds was discovered, and two days later another 4.15-metre male in nearby Struisbaai. Again it had wounds to its pectoral fins, and its entire squalene-oil-rich liver was missing.* All these sharks were identified as having been regular visitors to Dyer Island, and orca attack was listed in the necropsies as the most likely cause of the deaths.

Not surprisingly, talk among the shark diving community turned to how the orca could be driven away or even destroyed. More sharks washed up dead, and more weeks went by without any shark sightings. Through June and July 2017, only one shark was seen. Obviously it was no longer possible to run a wildlife viewing industry on stats like that. The internet lit up, and tourists stopped coming.

When my wife Helen and I stopped off in Struisbaai in 2018, it was to go whale watching and fat tyre biking over the sand dunes. Great white diving, we were told, was no more. The orcas have since been tracked heading all the way along this coast and northwest into Namibia. Wherever they go, the great whites vanish.

Marine biologists are keen to emphasise that this situation is more complex than just orcas killing sharks. There is no doubt that global shark fisheries are also decimating these highly migratory animals. A shark that was tagged here in South Africa swam all the way to Australia and back again in a single year; obviously this puts them at risk of being targeted outside the protected seas of Southern Africa. Also the local fishery for smaller species like soupfin sharks is decreasing the prey that

---

* Shark supremo Dr Mauricio Hoyos explained to me how he had seen orcas catching whale sharks by their pectoral fins, and pulling them like a Christmas cracker, creating neat rips along their flanks that allow them to get to their livers. Even more gruesomely, two orca in the Hebrides have been seen taking a grey seal by the flippers, swimming away from each other and ripping it in half lengthways …

young great whites rely on. And the protection of great whites is itself far from perfect; many great whites are caught on lines or in shark nets.

However, the orca story is a powerful one. I've seen many times the effect orca can have on their prey. When working in Alaska, we were filming a pod of common dolphins from above by helicopter. They were moving easily north, leaping and gambolling as they went. Then suddenly, on an unseen cue, the entire pod banked east and rocketed off at the surface, faster than it would seem a dolphin should be able to swim. Just minutes later a small pod of orcas were spotted cruising into the area. Everything fears the orca … even the great white.

It was 2022 when the first drone footage of the Holy Grail was secured: actual evidence of orcas killing great whites. Two animals attacked the shark from below, rammed it and then appeared to rip it apart. As with the Gansbaai sharks, all that was eaten was the liver.

· · ·

It was 2011 when I got my first chance to swim alongside a great white outside of a cage in blue water, in the only place in the world it can, in my opinion, be done in relative safety: Guadalupe Island off the Pacific coast of Baja California, Mexico. Guadalupe is a volcanic seamount, with no human inhabitants other than the odd biologist who might be ashore in a tent to study the annual arrival of elephant seals. Its more famous inhabitants are also drawn in by these leviathans, and by the endemic Guadalupe fur seal.

It's a fair jaunt out to Guadalupe, especially on the 'rustic' tub we were taking to get there; two full days and a night of steaming over empty and often tumultuous blue before the island appeared.

The crew for our adventure was the saltiest collection of salty old sea dogs ever assembled. Dive supervisor Richard Bull is a legend in the business, firstly because he was commercial diving back in the days when Deane helmets were still in use and surface supply hoses were held together with sticky tape and a prayer. Secondly because he had an alternative life as the bass guitarist for a band called the Kursaal Flyers who even made it onto *Top of the Pops* back in the seventies when that was a really big deal. Our dive medic was Chief Mike Hudson, a bear of a man with Mike Tyson neck folds, a wicked white goatee and arms that looked like he could bench press a bison. And my dive buddy was Mexican freediving supremo José Antonio Aguilar, never seen without a colourful bandana over his head, tough as a welded rivet, but the nicest person in the business.

The first thing José Antonio did when I stepped on board was to pull from beneath the boat's benches a chunk of rubber the size of a boogie board. It had a semi-circular chunk nearly a metre in diameter taken out of it. 'This was in 2008,' he said. 'A big female great white we called Bella kept on buzzing the Zodiacs. When she came alongside it, she was as big as the damn boat! Then one day, she didn't like the Zodiac, and just took a massive bite out of it.'

The final member of the team was not yet a legend, though he is now. Dr Mauricio Hoyos was just beginning to forge a reputation as the finest shark scientist in Mexico. Nowadays he's known as one of the best in the world.

'Bella is definitely big,' Mau said. 'And she's here this week. But we also had Deep Blue here a while back, and she is just amazing. We have a drone shot looking down at her and she was the same size as our tender, and that's near seven metres. I mean she was at least 6.1.'

'She is not as scary though,' José Antonio said, cutting in. 'I mean at that size she's so slow, kind of lumbering. She bangs into the cage and cuts herself up. She's almost clumsy.'

'So actually the bigger the better?' I asked.

'Sometimes, but you've got to read the animal,' Mauricio replied. 'Watch their body language. Are they moving quick? Is their mouth open and their gills flaring? Then no, you stay inside the cage. Is their attention on the bait? Are their pectorals down? Then they are getting ready to make a tight turn, so you stay in the cage.'

'But when you go out, you have to mean it,' José Antonio said. 'You commit, you swim out, and then you have to own the water.' He puffed out his enormous barrel chest and drew out his arms, like a big guy who was trying to look as intimidating as possible. 'The shark is after the weaklings, the ones who are injured, too old or too young. You look big and strong in the water and they're not interested. And they like to sneak up on you when you're not looking, so keep your eyes everywhere.'

'That's right,' said Mau. 'It's the shark you don't see that's the problem.'

'You have to own the water,' José said, emphasising the point. 'I'll always be watching your back, you keep your eyes looking forward and this will be cool. Trust me.'

When we moored at Guadalupe, there was another boat there already. They'd been filming for Discovery's Shark Week, and had been chumming the waters for several days. We hadn't even dropped anchor before the first great white came cruising around our back deck. One of the concerns about commercial shark diving is that chumming makes sharks associate boats with food. At long-standing shark dive sites like this, just the noise of your boat engine is enough to draw sharks in from a good distance away.

Already we had three darting immature males, unsure and constantly alert for the presence of larger animals that might take exception to them being around. While they had a lot of growing to do, they were still three metres in length and perfectly capable of biting a human clean in half. I have to admit my guts were tangled. The goal of swimming in blue water alongside a great white was to prove to the world that these animals do not mindlessly target humans and are actually incredibly selective about their prey, but now I was here and there were so many sharks, it didn't seem such a smart idea any more.

Our first few days of dives were all from the safety of the cage, to enable us to get a feel for the sharks and the conditions. The cage was sunk down four metres below the surface, so there was a nervous moment as you entered the water without protection and sank down into it from above, but in seconds you were within its prison bars and could watch the spectacle outside them unfold.

There's something otherworldly about the first approach of a great white towards you. It's one of the most iconic images in nature – something every human has probably seen on screen – and yet somehow it still seems unreal. How can any animal be approaching so slowly and yet with such menace? How can an animal that big be invisible in completely clear water, until it is almost upon you? The mouth is open, those famous teeth are on display ... how can you say or think anything new about this, one of the most familiar views in our oceans?

Those first few days taught me some very valuable lessons about great whites. Simply put, in the presence of food, size really does matter. The smaller animals only stick around until something bigger turns up, then they scarper. If two larger animals are approaching the same food, first they'll swim towards each other, and then one will turn so they are parallel and can assess each other's size in the most literal way – like two kids in a classroom standing back to back to figure out who's tallest. The bigger animal commands the food, the smaller beats a speedy retreat. One female was particularly noticeable, as she had characteristic scarring to her gills, great circular flaps over the slits where a male had taken hold of her during mating. She appeared to be the most relaxed animal of them all. José took my arm and pointed to her, then nodded, and gave me the universal diver's symbol for OK – that's our girl, he seemed to be saying.

After a few dives, we graduated to sitting up on top of the cage. Here you are exposed to the big blue on all sides, but beneath you is the protection of the bars. As great whites will often investigate food from beneath, this is a good step up, and allows you above all else to build your confidence. It also enables you to truly appreciate quite how big a role vision plays in these calm, clear waters. At first glance you could think the waters empty of life, but get your eye in, look a bit harder and there might be as many as seven or eight great whites within your field of vision.

We positioned ourselves at all compass points of the cage top, to ensure we had every angle covered. Towards the end of one dive, a big shark lunged at the surface towards the bait line, then dropped back into the water and kept swimming – right over the top of the cage, scattering me and the other divers like skittles.

Prey animals very clearly walk the line of visibility and perception too. For the entire duration of one dive, a particularly chunky female shark had a tiny juvenile Guadalupe fur seal just dancing around its snout, mere inches from the mightiest jaws on earth. The seal did somersaults, twerked its backside and feigned toothy snaps towards the huge daunting animal, obviously completely

unconcerned about its reputation. This is the equivalent of ante-
lope pronging or stotting (performing straight-legged leaps right
in front of prides of lions) on the African plains. It's what's known
in biology as an honest signal – a statement of fitness that cannot
be faked. It shows the predator that they should not waste their
energy in a pursuit that will be fruitless; the smaller animal sees
them and possesses manoeuvrability and skills they cannot hope
to better. This seal's display must have been an honest signal,
though it seemed brazen or even foolhardy to perform in the face
of a giant white.

The great whites won't stray into the shallowest waters right
next to the shore, and here you can drop into the finest aquarium on the planet without having to keep your wits constantly
about you. The waters are so clear you can barely tell you are
submerged, and it almost feels like flying. The fur seals are
some of the most interactive you'll ever see, twirling about
you, looking at their reflections in the glass of your mask or
camera. They are always on the lookout, though. Unlike most
seal species near their colony, if the animals are resting, it is with
their rear flippers out of the water, and their bodies orientated
vertically, eyes always looking downwards and outwards, alert
for the threat of the shark.

On the evening of our third day, the team made the call: the
following morning would be our opportunity to swim outside of
the cage with a great white. There was no pressure on me to do
so. I had to be doing this of my own accord and because I felt it
could be done safely.

The final morning dawned on a perfect Guadalupe day. Flat
calm seas, blue, blue water, and lots and lots of sharks. It was
now or never. As so often in these scenarios, the team and I were
quiet over breakfast. I then set to cleaning and checking the
talking mask system I use when diving, checking and rechecking

all the valves and hoses. Burying myself in simple tasks is the best way of diverting myself from nerves. I might not be great at conquering my fears, but I am exceptional at ignoring them.

Every member of the crew took me aside that morning and quietly told me that I didn't have to do this if I didn't want to. The divers because they needed it to be my decision, not something I felt I was being forced into. The non-divers because they'd seen the sharks from the boat and were frankly certain I'd get eaten. But by then I had made up my mind. We just needed the right shark.

This time we dropped down to the safety of the cage, but instead of climbing up on top, we rolled back the bars at the front of the cage and crouched ready in the doorway. I had a 15-litre steel cylinder on, with a further 6-litre 'bail-out' cylinder strapped to the side of it. At five metres in depth, where we were now, I could wait for more than two hours before I'd need to return to the surface for fresh tanks. We could bide our time and make sure everything was right.

For the first hour the water was eerily quiet, but then a familiar shape loomed. With the circular scarring around the gills, she was identifiable as our placid female Bella. José grasped my upper arm and motioned out into the water. I swam out. Our female totally ignored me, swimming just metres over my head, alongside and below me. At one point I broke the golden rule and ended up in between her and the bait. She swam straight at me, then at the last moment detoured and swam around me to get to the food on the other side. We had the entirety of our second hour swimming in total tranquility alongside our big girl, and though it started as a nervous game, by the end it had truly been the most life-changing animal encounter of my life.

Not everyone saw it that way. Marine biologists and conservationists I respect have said that if I had been bitten, the sharks would have been vilified and their malign reputation exacerbated. Ultimately it's the sharks that would suffer. People have said that it is unacceptable vanity and recklessness to risk so much for such little payoff. Perhaps they are right.

Now that I have taken the ultimate step to prove that a shark's danger to us is overstated, would I allow my own kids to dive with them?* That's difficult, because I wouldn't ever want to expose my children to risks, even if those risks are small. The question is even more difficult to answer because my understanding of shark attacks has been complicated by nuance in my own experiences. Ten years ago, when I had done maybe a thousand shark dives, I thought I knew everything and that sharks were no more dangerous to us than cockapoos. But that is only true of most shark species most of the time. In addition, you are much safer as a big, strong, healthy, confident human. Like all apex predators, sharks pick their targets.

Mauricio tells a tale that exemplifies this. A small group of wealthy divers turned up at Guadalupe with the goal of swimming outside the cage with great whites. However, as soon as he saw them jostling for position in the cage, Mau could tell that one of them – a German man in his thirties – was scared. Mau signed to the dive master with the universal symbol of a hand held horizontally and wobbled side to side. 'So so,' he intimated, and pointed at the German. 'He's going to be a problem.' But the tourists had travelled a long way and spent a lot of money to have this experience. They insisted on continuing and swam out into the water alongside the sharks.

Mau described an instantaneous change in behaviour from the predators. They normally cruise by disinterested, but this

---

* Here I mean the big predatory species like bulls and tigers. By three years of age, my kids had already swum with several kinds of reef shark, nurse sharks, mantas, eagles and stingrays!

time they turned, pectoral fins dropped and backs arched, and headed straight for the German, with clear predatory intent. They sensed he was afraid* and that fear made him potential prey. The tourist was so frightened that he grabbed his own dive master and thrust him in front of the incoming jaws. Mau says they were lucky to escape without a fatality.

There are so many other complicating factors in human–shark interactions. Mau has published a paper showing that the type of bait in the water elicits different behaviour from the sharks. Fresh tuna blood, for example, brings in the same number of sharks as those brought in by old chum, but seems to drive those that are around into a frenzy. Time of day and visibility also play a part. I'll never forget spending a week with tiger sharks in a Caribbean hotspot. We dived all day every day with them, got to know each animal by name, thought of them as being like big old puppy dogs (despite them being able to bite a turtle in half, shell and all). On the last day, we thought we'd chance getting in at dusk. They were completely different animals. The second we entered the water they swam straight at us, hard and fast, their formidable teeth visible in their gaping mouths. We had to physically bump them off with our cameras and it was less than five minutes before we rushed for the surface with our tails between our legs.

All sorts of things can trigger a change in behaviour. We've noticed that big animals bothered by hooks and lines that dangle from their mouths can be belligerent and sneaky. Perhaps the hooks compromise their natural feeding and they are forced to be more ruthless? Instagram videos show free-divers calmly 'redirecting' goofy-looking tiger sharks away as

---

* As I will come back to in Chapter Eight, signs of stress include elevated heart rate and breathing. Transmitted through the water, these are the exact signals the sharks focus on.

they nuzzle in to check them out, and this technique absolutely works. With a standard tiger. However I have (just once in 25 years) seen a five-metre tiger hitting an object at the surface with true intent. It swam up like a great white breaching on a seal at more than 20mph. Nobody would be gently nudging away a shark doing that.

And I have been bitten by sharks several times. Once was while filming three-metre lemon sharks and hand-feeding them. I was surrounded by possibly 40 sharks, to the extent that you couldn't see me within the swarm of animals, but they never tried to bite me. They knew what the fishy food was and weren't interested in me. However, I was also trying to lure the sharks into a *Matrix*-style time slice camera rig to bite a fish, so we could show the way a shark's bite works. One shark took not just the fish I was offering it but my whole hand, and swam off with me into the blue. If I hadn't been wearing chain mail, I would have lost my hand.

There is a lot of debate about whether shark attacks are falling (due to a fall in global shark numbers) or if they are actually on the rise. The Australian Institute of Marine Science says: 'The rise in Australian shark attacks, from an average of 6.5 incidents per year in 1990–2000, to 15 incidents per year over the past decade, coincides with an increasing human population, more people visiting beaches, a rise in the popularity of water-based fitness and recreational activities and people accessing previously isolated coastal areas. The majority of incidents were in the warmer months November to April.'

There is no evidence of increasing shark numbers that would influence the rise of attacks. The risk of a fatality from shark attack in Australia remains low, with an average of 1.1 fatalities per year over the past 20 years. There has actually been a decrease in the average annual fatality rate over the years, falling from a peak of 3.4 per year in the 1930s. This means that fewer people who get bitten now die, which correlates to an improvement in primary care for trauma victims, combined with a decrease in response and evacuation time.

The risks of shark attack are undeniably very small, but – and it's a massive but – they are not zero. My life is one big risk assessment. Doing what I do for a living, I spend my waking hours weighing up the potential of misadventure against the steps you need to take to minimise the danger of the worst possible outcome occurring. The chances that a shark is going to bite you are low, but any bite from some species of shark *could* kill you.

So what can be done to reduce the risk? Well, the website for the Australian Institute of Marine Science (AIMS) has lots of info on the places in Aus with the highest incidence of shark encounter, and the months when they are most likely. A little knowledge about the behaviour patterns of sharks also helps. The weeks we've spent filming great white sharks launching breach attacks make it clear they hunt like this at dusk and dawn, so it makes sense not to swim at these times, and to stay clear of seal colonies and murky waters. It's also wise to avoid being in the water around estuaries, piers and harbours, especially where fisherman are cleaning their catch. Spearfishing and other kinds of diving where you're collecting seafood are definitely risky.

To suggest people should steer clear of the sea in certain conditions probably sounds crazy to people living in European countries with much tamer waters. However in places like Australia, Brazil and certain parts of the States, many people have a different attitude to wildlife. Australia is a nation that has learned to live alongside the world's largest crocodile and most venomous creature (the box jellyfish), not to mention its roll call of the world's most toxic snakes and spiders. People in many parts of America have to deal with alligators, bears, rattlesnakes, cottonmouths and mountain lions. These animals cause only a handful of fatalities, because people have learned the rules of living alongside them in harmony. People are vigilant near rivers in the Northern Territory in Australia, don't swim in box jelly season, don't touch any snake unless they're 100 per cent sure what it is. Despite the array of venomous spiders, due to understanding and antivenins no one Down Under's been killed by a spider in 35 years. Risk management and mitigation is a way of life in these places, and it works.

So beyond sensible behaviour and deepening our understanding, what else can be done to mitigate human–shark conflict? Scientists point out that education and awareness are more effective than culls. Great whites are not territorial; rather they're pelagic fish that are always on the move. *Jaws* was a pure fiction; you don't get great whites that develop a taste for human flesh and hang around in one place munching them. Drum lines (floating unmanned baited hook traps) are utterly indiscriminate and kill a huge range of non-target species, from turtles to dolphins. Culls kill hundreds of sharks that are just minding their own business on their way to somewhere else and would never harm a human. And let's not forget, great whites are vulnerable and protected, so no one should be able to just kill them for the sake of it. There should be bucketloads of solid science proving culls work before governments are allowed to just wade in and start killing them.

. . .

The most important reason that we cannot allow sharks to be persecuted as a result of the vanishingly few attacks on humans is that we are on the brink of eradicating some shark species already with uncontrolled fishing.

At least 75 million sharks are targeted annually for their fins alone, which are bound for the Chinese wedding delicacy of shark fin soup. However, it is dangerous to blame all the shark's woes on overconsumption in the Far East. That othering attitude suggests that the decimation of sharks is the fault of people on the other side of the world, and implies we in the West have no culpability. This couldn't be further from the truth. The biggest fisheries for sharks are actually from the European nations of Portugal and Spain. And, as I've already mentioned, threshers are found and caught right here in British waters.

Many people are surprised to hear that we have perhaps 50 species of sharks in British seas, and that even sharks as exotic as the scalloped hammerhead have been recorded here (one washed up on a Dorset beach in the 1880s and is now in the Natural History Museum's vaults). The nearest thing we have to a great white is the porbeagle shark. They look like a smaller version of the great white, feeding on squid and other fish and sharks, and they share some of the vulnerabilities of their more iconic cousin. They're long-lived but also take a long time to mature, pregnancies last for perhaps a year or more, and they produce few young. It's a common strategy among sharks and other apex marine predators that has worked fabulously for them over the last 440 million years. That is until they were faced with the most efficient hunter this planet has ever seen – us.

Porbeagles have been fished to the brink of extinction, mostly for their meat, oil and fins, which are again used in shark fin soup, and as accidental bycatch by fishermen searching for more commercially viable species. They are often sold as 'sword-fish', which the meat resembles. Now critically endangered in the North Atlantic, we face decades of panic management to make sure paltry populations of this wonderful shark stand any chance of surviving. And they are not alone; angelshark and common skate are functionally extinct in our waters, and overfishing of tope and spiny dogfish (once our most abundant sharks) has meant there is simply no point in commercial fishermen trying to catch them any more.

It's an environmental nightmare that has become the norm in recent years. We don't recognise that habitats are valuable and animals are close to extinction until way past zero hour. Too often we in conservation cheer when an animal is listed on CITES,* a global recognition that the animal is endangered, because this enables trade bans and sparks the awareness needed to start protecting these species. Surely it would make much more sense to recognise which species are vulnerable early and take steps to limit their exploitation in order to avoid expensive and often doomed 'panic conservation' in the future?

Sharks' biology combined with our fishing practices make them peculiarly vulnerable. These are species that are vital to the wider ecosystem health. They control the behaviour and distribution of potential prey animals just through their intimidating presence – the presence of tiger sharks, for example, has been shown to stop turtles overgrazing seagrass beds. Decimation of sharks has in some places led to an increase in the rays they would normally feed on, which in turn leads to those rays overeating their own chosen prey, which include food fish like scallops, abalone and oysters.

Research tells us what we can and can't catch. There just needs to be regulation put in place to make sure we keep our catches sustainable. History shows us that unrestricted exploitation leads to populations crashing, potentially never to recover. We need to act to stop uncontrolled shark fishing now, adopting effective management before crisis recovery plans are needed.

---

* The Convention on International Trade in the Endangered Species of Wild Fauna and Flora.

The high seas can no longer be an out of sight, out of mind wild frontier where anything goes.

We could lose many species of sharks from our seas within my lifetime, and that simply must not happen. Lose sharks like the great white, the mighty, mysterious lords of the deep, and our planet's oceans would be infinitely poorer places.

CHAPTER EIGHT

In Cold Blood

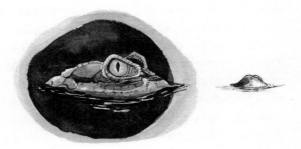

All of us have in our veins the exact same percentage
of salt in our blood that exists in the ocean, and,
therefore, we have salt in our blood, in our sweat,
in our tears. We are tied to the ocean. And when we
go back to the sea – whether it is to sail or to watch it –
we are going back from whence we came.

**John F. Kennedy, remarks at the dinner
for the America's Cup crews, 14 September 1962**

*'Niño!' our guide calls out with the tone of someone calling a golden
retriever to have its tea: 'Here, boy, where are you?' He slaps a piece
of puckered chicken thigh on the water, and calls out into the Cuban
mangroves, 'Niño!' He has been calling like this for perhaps half an
hour. Below us, clear, shallow seawater reveals luxuriant seagrass
meadows. Of all the strange ways of attracting an animal, this has
to be the strangest. And then our boat is stirred into action; there it*

is! Making a beeline towards us through the twisted mangrove roots is a profile that strikes fear into any human with a heartbeat. We can see nothing more than the nostrils and ears, then the twinned scales running down its massive tail. Like an iceberg, a croc's bulk and lethal potential are all below the surface. Every instinct in your body is to recoil, but our guide is gesturing at us to get ready. I step to the other side of the boat and get in – and sink straight up to my knees in seagrass mud, which swirls around until I can't see the bottom any longer. This cannot be a good idea.

I put my face in the water, willing the murk to settle. As it does, I recognise a familiar slit-shaped pupil and a dragon-like snout with white teeth spilling out of gnarled jaws. The crocodile is lying right in my lap, a hand's breadth away from my most precious possessions. A single sideways snap and any chance of a Backshall dynasty would be gone forever ...

• • •

Marine reptiles are a curious group. They are comparatively uncommon; of perhaps 12,000 species of reptile, maybe a hundred live in marine environments, and of those only some sea snakes truly live out their whole life cycle at sea without coming to land. With saltwater crocodiles and blunt-snouted marine iguanas the most dinosaur-like creatures alive on earth today, and turtles tracing their lineage back to before the dinosaurs walked the earth, they seem to be the most ancient of all life forms, and are often described as living fossils or evolutionary throwbacks. However, marine reptiles are all versions of animals that left the sea, evolved into something close to their current form, and then returned, making them secondarily adapted to aquatic or semi-aquatic lives.

The earliest marine reptile may have been *Mesosaurus* in the Permian period, perhaps 299 million years ago. The more familiar ichthyosaurs and plesiosaurs were to follow. Some of these creatures became so adapted to the sea that they never returned to land, even giving birth in the water. Most went extinct with the dinosaurs at the end of the Cretaceous. This leads to some interesting questions. Why do we still have marine reptiles today? How did they survive when so many other creatures of their kind perished? Why did the dinosaurs die out but the crocodile did not?

The answer is that crocodiles did die out with the dinosaurs – just not all of them! The story of the crocodile is one of the most fascinating in palaeontology. Appearing in the fossil record at around the same time as the dinosaurs, there was a far greater diversity of crocodilians around during dino times, and not all were like the ones we have today. Some grew to a huge size; *Dinosuchus* and *Sarcosuchus* could have fed on some of the largest dinosaurs. However there were also small, lightning-quick insectivore and dedicated fish-feeding crocs, and fossils show us there were crocs that were herbivorous, as well as serpent-like marine crocs and terrestrial crocodiles – ones that could climb trees, run and chase down dinosaurs on land and would rarely or never have gone into the water.* They were adapted and specialised to a wide variety of different niches.

Some were almost identical to those we have today, and they, like the turtles, would have benefited hugely from their amphibious life cycle. When the water became de-oxygenated or overheated, they could move on to land. When the plants died and the land was scorched or frozen, they could shift into the sea.

At first sight, crocodilians would seem to be similar to dinosaurs, but they are totally different. The first of those differences is in their metabolisms. Crocs are so-called 'cold-blooded' – in

---

* Indeed *Mesosuchus* was a modern crocodilian that was terrestrial, and only went extinct from its Pacific Island home around 3,000 years ago – funnily enough, shortly after humans arrived there.

scientific terms ectothermic, meaning they get their warmth from outside sources such as the sun rather than from consuming food, and poikilothermic, meaning they track the temperature of their surroundings. Current thinking is that the closest relatives of dinosaurs are modern birds (now known by some taxonomists as avian dinosaurs) and that dinos shared more in common with birds than with living reptiles. Most groups of dinosaurs were at least partially feathered and almost certainly warm-blooded. Warm-blooded or endothermic animals thermoregulate by consuming calories and burning those internally through metabolic function. Being warm-blooded has huge benefits, as it allows animals to keep functioning even when they are cold. There are also significant downsides though, in that endotherms require huge amounts of calories and constantly need to refuel. Very few modern warm-blooded animals can go without food for more than a few days, and it is probable that dinosaurs were the same.

The asteroid that struck the Yucatan at the end of the Cretaceous did not wipe out all life forms (except presumably at Ground Zero in Central America). Life was choked out over years or even decades, as the sun was clouded with haze, plants died off and the large dino-herbivores faded away. However, in extremis, crocs can go into a sort of stasis and not eat for as many as three years. Studying the genome of crocodilians suggests that their genetic evolutionary change was rapid in favourable times but slowed to nothing during challenging times. An animal that could gorge on the corpses of dying animals, then go into a burrow (as modern alligators do today) for years on end with no ill-effect was uniquely placed to succeed and survive.*

* * *

* The few mammals survived because those around at the time were tiny, and could hide away and scrape by on small amounts of calories.

It may seem strange to include crocs in a chapter on marine reptiles. After all, of the 23 species around today, only the American and saltwater crocs routinely go far out to sea. They are, however, two of the most successful of all croc species, with the saltie the most widely distributed – it can be found from New Guinea to India and lots of other places in between. Their ability to swim out to sea and on to different islands is probably the key to their broad distribution, and their ability to tolerate salt or fresh water is central to their success. Osmoregulation (the ability to regulate dehydrating saltwater) is their greatest challenge, solved using salt-excreting glands in the mouth. They're also the biggest of the crocs. The longest on record was measured by Dr Adam Britton in the Philippines in 2013. Named Lolong, he was 6.17 metres in length and weighed more than a metric tonne!

Their ability to adapt has been a factor in many of my crocodile encounters over the years, including what I consider to be the closest I – or anyone on my team – has come to being killed by a wild animal. In 2011, Britton had been working on DNA analysis of crocs in the Okavango Delta in Botswana and had put me in touch with legendary wildlife camera operator and adventurer Brad Bestelink. Brad had spent much of his life scuba diving in the waters of the Okavango, and on one occasion he had accidentally ended up swimming with his wife Andy into a giant Nile croc on the bottom of the swamp. Instead of the croc going for them, it had totally ignored them, leading the couple to start diving with the animals deliberately.

Six years and hundreds of dives later, Brad had developed a theory that it could, in certain circumstances, be safe to dive with Nile crocodiles. This seemed to me to be utter lunacy. However, Brad reported that in the Botswana winter the water is so chilled in the Okavango that the crocs become sluggish and simply do not move.

Brad was the most experienced croc diver in the world, and crocs were part of his family history too: his grandfather was a croc hunter estimated to have killed 30,000 crocs. He wooed me with extraordinary footage of animals lying motionless in sunken caves of papyrus, totally ignoring him as he came to within inches of them. Plus he had taken Anderson Cooper, of *60 Minutes* fame, and Ben Fogle out diving with him, neither of whom had much dive experience at that point. Surely it should be possible for us?

So in February 2013, we headed out to the most remote corner of the delta, taking all our dive kit with us. We had ten days to try to document the crocs underwater. It was a simple but intense way of working. We'd power down the smaller waterways of the delta, binoculars constantly scanning for the presence of a croc resting at the surface. There were few areas of dry land where the crocs could bask, the majority of the environment above the surface being floating lilies, reeds, swamp and the endless papyrus. The only areas where animals could come onto land were precious and tended to be dominated by bigger males.

Before we made our first dive, Brad gave us an extended and frankly terrifying briefing. 'The surface is the kill zone,' he said in a matter-of-fact tone. 'Animals struggling silhouetted at the surface are a target for crocs, so don't let that be you.'

'It's all about getting down to the bottom as quickly as possible,' said his wife Andy.

Both Brad and Andy's skins were tanned walnut brown from a lifetime of African sun, and they had black hair and dark eyes. They were a wild, tough and imposing couple. Brad was wearing an old-fashioned rubber wetsuit top with no sleeves, showing off his significant biceps. He reminded me of Burt Reynolds' tough-guy character in *Deliverance*.

'So empty your BCD* of all air,' he continued, 'put on more weight than you're comfortable with and just drop like a stone.'

'What if you have trouble equalising?'† asked (not unreasonably) my Belfast-born cameraman Johnny Rogers – himself a child of the Troubles who had seen pretty much everything.

'Then I guess you burst your eardrums,' said Brad. I wasn't sure if he was joking.

'What's the situation with bailout cylinders and talking masks?' asked Richard, the dive supervisor.

'No bailouts,' said Brad. 'If you're in trouble you get up and out of the water, and you're not going to want any extra weight holding you back.'

'And talking masks … you can wear one, but be careful about talking near the crocs,' said Andy. 'I'm not sure they'll like it!'

'When you're underwater it's all about being dominant,' said Brad. 'You need to be big in the water, move slowly and calmly, but always with confidence. Don't let them dominate you, or you'll be in big trouble.'

This I understood. Whereas we're taught to be submissive around certain animals, such as gorillas, I had of course been told the exact opposite about great whites. Exude confidence. Own the water.

'Crocs have these special organs on their lips,' Brad continued. 'They are really good at picking up vibrations in the water. So let's say an egret lands, and it's injured, it's struggling, it's sending out these little ripples into the water that give it away. Don't let that be you.'

Just like when diving with sharks, it would be critical to stay calm and relaxed. If your heart rate and breathing went up

---

* Buoyancy control device – what non-scuba divers would probably call a life jacket!
† 'Popping' your ears to equalise the pressure.

because you were stressed or scared, you'd be sending out those signals into the water, and the predator would detect them. It would flip a switch in their reptile brain, and all of a sudden they'd be in kill mode.

The first time we dropped into the water was the most sinister experience I've ever had. The environment was like something out of a Hollywood horror movie. The waterways that from above shone with flowering lilies under blue skies were beneath the surface dark and foreboding places. The Okavango River doesn't reach the sea; instead, it floods seasonally and flows out into an inland delta. When it does flood, the waters carry everything in front of it until the flow slows and it drops its cargo. The bottom, a silty brown sand, is therefore filled with dead trees of varying ages – some of them uprooted and still bearing brown leaves, others ancient skeletons. The papyrus grows into matts at the surface, but creates caverns beneath. Everywhere rotting vegetation gathers in dark corners. The visibility is poor, and you often cannot see more than a few metres in front of you. There are a seemingly infinite number of places where a croc could be lurking.

We sank to the bottom and allowed ourselves to be carried along by the flow. I felt a tight grip on my upper arm, and instantly flinched, ready to punch the attacking croc in the face. It was just Brad's meaty hand clasping me. He was pointing toward one particular cave. Shards of sunlight cut down through the gloom, and I could instantly see what he was pointing at. Two rows of gleaming white teeth, shining with what appeared to be other-wordly brightness.

At first I couldn't see the rest of the croc, but we pulled ourselves along the bottom, using our hands in the sand as anchors. When we were touching distance, all of a sudden the full animal was visible; it was perhaps three metres in length, so comfortably capable of eating me. But just as Brad said, it was totally motionless. The croc allowed me to sit at its side and talk about its features to the camera for a good ten minutes and never moved so much as its nictitating membrane (the third eyelid that covers its eye when underwater, like a swimmer's goggles).

A week later, we'd had 20 or so such encounters, and had been totally won over by Brad's big idea. The water was indeed cold, and our ectothermic targets very rarely moved at all (though some lazily swam away from us if we crowded them). They were predictable and of no danger to us whatsoever. We had not, though, filmed a giant croc, and that was what we had been sent here to do.

We were out late in the afternoon on the penultimate day when Brad abruptly slammed the boat into reverse, throwing us all to the deck. 'There!' he shouted, pointing into the lilies. 'Big croc, huge, just gone under!'

The crew geared up and in three minutes flat we were in the water. We'd become a well-oiled machine over that week, having done as many as 20 shallow and short dives each day. We all knew our jobs and the protocols and exactly how to work.

All of that came crashing down as soon as we saw the croc. The first thing to hit us was simply his size. I say his, because females don't get this big. He was at least four metres and proba- bly four and a half. This might not sound like much of an increase when the other crocs had been three metres, but a small increase in length can lead to an exponential increase in mass. This croc was huge, half a tonne at least, and capable of taking down a buffalo. And while the others had all ignored us or swum away, this big male stood up in a commanding position and swung his head in our direction, clearly aware that we were coming. Brad pushed me on towards it.

'Keep calm,' I kept saying to myself. 'Keep your heart rate and breathing down.' I dragged myself through the silt till I was sitting alongside its mighty tail. 'This is the closest I've ever been to lying alongside a dinosaur,' I said,* just as the croc swung its snout around towards me. I recoiled, but it turned and headed up to the surface, its tail sweeping past me just like a mythical dragon's. I looked to the others and shrugged as if to say, 'What a shame,' though I was actually relieved. Above me, though, the croc simply took a breath, turned and headed straight down and towards us. All three of us splintered and swam in different directions, and the big male came right through the middle of us, before disappearing into a cloud of silt. The exact same cloud of silt where Johnny the cameraman had just been. This was the most frightening moment of my whole career.

In a few seconds, my mind played out the complete scenario. Johnny had been taken. We would have to flee and would not be able to reclaim his body. I would have to go to see his partner, and his two little girls, and tell them their daddy was a hero but wouldn't be coming back. In that instant I felt the guilt and horror of having to do that, and of losing one of my oldest friends.

And then, just seconds later, the cloud cleared, and the croc came right out through the middle of it with Johnny mercifully evident behind him. The three of us could have reached out and run our hands down its dragon-scaled flanks. One sweep of its jaws to either side and we would all have been maimed or worse. We forgot our dive brief and all swam off in different directions, scrabbling our way up through the papyrus and reeds in utter panic, until the boat could come and rescue us.

That night, sat around the campfire, we digested what had happened. We tread a fine line with wildlife when we are on expeditions, but with the understanding that we know the animals,

---

* I cursed myself for this later, as all my biologist friends would delight in telling me he was nothing like a dinosaur!

know the conditions, and have the experience to make it safe. This time, though, we all knew that we were alive and unscarred through nothing more than pure blind luck.

'It wasn't cold,' Brad said. 'I guess a male that size gets to choose a basking spot on land. It must have been warming itself in the sun, so was good and warm and not sleepy.'

'And big animals hold heat better,' I added. 'Allen's rule and all that.'*

'So you think you'll carry on doing it?' asked Johnny finally. There was a long pause.

'Not me,' said Andy, who had been our safety diver up on the boat the whole time. 'I'm never diving with crocs ever again.'

As for me, I've dived with nine species of crocodile since, but never another even close to his size. None have ever showed me any aggression. Perhaps it exemplifies how, in the world of the crocodilian, size really is everything.

· · ·

Though modern marine reptiles are far less diverse than at many times during prehistory, we still count among their number many sea snakes and sea kraits, and of course the nine species of turtles. All of these evolved from terrestrial ancestors that evolved mechanisms for making the sea 'bearable'. In turtles, that is a form that looks much like a modern bearded dragon but with a protective shell and paddle-like limbs. They also developed modified lachrymal glands to enable them to 'cry' excess salt out of their bodies, and all turtles return to the surface to breathe oxygen (though some can undergo cloacal respiration –

---

* Allen's rule states that while an object doubles in size, its volume increases to the power of three, while its surface area only increases to the power of two. Simply put, a larger animal has proportionately less surface area to lose heat through. It explains why large animals are much more effective at regulating heat than small ones.

that is, breathing through their anuses) and to land to give birth to their young.

The turtle body plan arose in the Triassic, as the earliest dinosaurs were making their way across the planet, and has remained remarkably unchanged in the 250 million years since. There are also many examples, thanks to convergent evolution, of animals that were superficially similar, in that they were relatively slow-moving and long-lived, and protected themselves with a heavy shell; for example, the *Ankylosaurus* among the dinosaurs, and Pleistocene mammals such as the *Glyptodon*. Placodonts, which were a family of reptiles that also cropped up in the Triassic at about the same time as turtles, appeared to spend their time both on land and in the water, and with their stubby legs and tough beaks were similar in appearance to the early turtles. However, they became extinct 200 million years ago, and are not, bizarrely, an ancestor of the modern turtle. Instead, the ancestor of turtles appears to be *Eunotosaurus*, a reptile that is first seen in rock deposits dating to the Permian. This shelled creature was purely terrestrial, more like a giant tortoise you might find in Mauritius or Galapagos nowadays.

In 2008 in China, palaeontologists discovered *Odontochelys*, which appears to have swum in Asian seas around 220 million years ago, and *Proganochelys,* from Europe, has been dated to 10 million years later. This oddity had a clubbed tail like *Ankylosaurus,* which it used to protect itself, spikes on its neck (so it couldn't have retracted its head into its shell) and teeth inside its beak. *Odontochelys* had teeth on both upper and lower jaws, and lacked a fully developed carapace to the shell. That carapace is a big part of the success of the modern turtle group, being formed by the fusion of the vertebrae and the ribs to form interlocking protective plates.

By the early Jurassic, 200 million years ago, fossils of both tortoises and turtles look very much like what we have today. In the Cretaceous, we get the giant *Archelon*, which would have been much like a modern leatherback, but twice the size – four metres in length across the carapace of the shell and weighing two tonnes. How the females dragged themselves ashore to drop their eggs is anyone's guess.

. . .

To lie silently in pitch blackness alongside a turtle as she lays her eggs is one of the most primordial experiences imaginable in the modern world. In Trinidad in 2014, we used thermal imaging and infra-red to spy on giant leatherback females as they made their nests in order not to disturb them from their tasks. This was a very welcome improvement on the very first time I saw leatherbacks nesting in Malaysia's Rantau Abang in the mid-1980s. Then huge crowds with giant lights gathered around every turtle, and people posed for photos riding on their backs as they laid their eggs. Today we understand far better how precious these moments are to turtles, and how easily they can be thrown off their stride. Many turtle beaches around the world are well protected, and artificial lights are dimmed or even completely banned.

In Trinidad, great tank tracks ran up and down the beaches, where the half-tonne females had dragged their bodies over the sand in order to lay their eggs. They might have travelled 10,000 miles to return here to the same beach where they themselves hatched out decades before. There was a real sense of ritual about our ancient female's return. She lumbered out of the waves, milky-eyed, pink and salt-encrusted, like a creature from David Lynch's *Dune*, and in a torturously slow march dragged her belly up the sand to above the high tide mark.

She then prepared a spot with her foreflippers, before turning around and churning out a deep hole with her rear flippers. It was clearly exhausting work. Every few minutes she stopped, threw her head back and heaved with the effort. Salty tears ran from her ancient, rheumy eyes.

The eggs, when she lay them, were like mucus-covered ping pong balls (though slightly more leathery and squishy) – most turtle species lay around 60 of them. Then she piled sand back onto the nest and left.

Both turtles and crocodilians use this same technique, and both also exhibit temperature-dependent sex determination. This means that rather than the foetus's sex being defined by its chromosomes, the ambient temperature experienced by the embryos in the middle third of their development will determine the sex of the offspring. Only a very small constant range of temperatures will allow for a brood resulting in offspring of both sexes. Temperatures just a degree or so higher or lower than this standard will result in the brood being entirely male or entirely female. In the European pond turtle, temperatures below 25°C produce all males, and above 30° all female. You only get a mix at 28.5°! The eggs of the snapping turtle are all female at or below 22° and above 28°. Between these extremes, males are most common. It's uncertain what evolutionary advantage this phenomenon has provided these reptiles. Additionally, though reptiles generally undergo sexual reproduction, there are many cases of female reptiles giving birth in the absence of males. In biology, this is known as parthenogenesis. In the language of *Jurassic Park*, 'Life finds a way.'

• • •

To see one single turtle come ashore to lay is an experience to be cherished. However, in the northwest of the Pacific coast

While whales and dolphins usually sleep on the go – shutting down half their brain at a time – sometimes you find them hanging in the water, properly snoozing.

LEFT: A play date with a sperm whale is a full body experience. You can feel their acoustic powers vibrating through your whole body, as they assess you with sound.

BELOW LEFT: There is no more primeval experience than witnessing a crocodile underwater. I have dived with nine different species, and only once felt in mortal danger!

BELOW RIGHT: As you kayak alongside a fresh berg you can hear them creaking and groaning, threatening to collapse or roll at any second.

ABOVE: There is no greater gladiatorial contest in nature than the jousting of elephant seals, the sumo wrestlers of the marine world.

LEFT: Known as red devils, Humboldt squid have been known to travel in squadrons, and to cannibalise their compatriots.

ABOVE: This wandering albatross will have spent many years out at sea before returning here to South Georgia to nest. With the biggest wingspan of any bird, they live a life aloft.

BELOW: Lying in the sands of a South Georgian beach, the orchestra of sounds from the king penguins is at turns soothing and melodic – eventually becoming endless and sometimes driving you demented!

ABOVE: In the shark sanctuary of the Bahamas, it is sometimes difficult to find a shark-free bit of water to get into!

BELOW: Oceanic whitetip sharks are one of the species hit hardest by longline fishing. Once the most numerous and widespread large predator on earth, in some places they have disappeared completely.

This big tiger shark was an unexpected and welcome guest as we were boarding our boat. Curious and pugnacious, it hung around for hours.

ABOVE: Bubblenetting humpbacks have been described as the greatest spectacle in nature. Getting stuck right in the middle of the bubble net ... less than ideal!

BELOW: New technology is offering us a fresh window on marine environments, but when it comes to the next generation, sometimes the old-fashioned ways are the best!

of Costa Rica, hundreds of thousands of marine reptiles pour ashore, in one of the finest wildlife events on the planet.

Ostional is a six-kilometre stretch of black sand beach that plays host to one of the most dramatic, most enormous, most jaw-dropping spectacles in nature. For a few months of the year, and for a couple of days synchronised with the coming of the new moon, Ostional plays home to the greatest aggregation of reptiles. Olive ridley turtles – giant ancient mariners, their heads encrusted with barnacles and shells bearing the scars of shark bites and boat strikes – congregate at Ostional to breed, and soon after the females come ashore to lay their eggs. But unlike any other species of turtle, they don't sneak ashore silently and solo, hoping to evade the predators that would snatch their eggs and hatchlings. Instead they come ashore in multitudes so awe-inspiring that the predators are simply overwhelmed. In one 72-hour period on this one small stretch of beach, half a million have been counted coming ashore. The sight of it is just impossible to convey.

The peak of *arribadas* – arrival of the turtles – occurs at night and at high tide. The cameraman and I arrived mid-afternoon at low tide, so I didn't expect to find anything. Instead, as we had not slept on our journey from the UK, it seemed a long lunch and a short kip was in order. Not to be. We went down to the beach and were blown away by the scene before us.

Ostional is pretty impressive even on a normal day. Waves crash down on black sands, and jagged volcanic rocks just beyond the break send plumes of water like geysers into the sky. Behind the beach is uninterrupted rainforest, cloaking hills and then mountains. But now, all the eye could see was turtles, scrambling over each other's shells in their desperate mission to find a spare metre of sand to dig their nests, and battling up the gradient and colliding with their fellows heading back out to sea. Behind them in the breakers, countless heads popped up from the foam in a holding pattern awaiting their turn. There were 12-year-olds on their first nesting missions, with shells about two and a half feet along their axes, through to wizened octogenarians, with heads the size of coconuts and shells like a car door.

Standing in the surf, they washed in with the surge and knocked sharply against my ankles. Some were flipped by the waves onto their backs and flapped pathetically till they were righted. Further up the beach, all that was visible off into the distance was a sea of wobbling boulders chucking sand up behind them as they dug. Huge gusts of sand were propelled forcibly into my pockets by their vigorous digging. I couldn't walk more than a pace without stumbling over a flipper. We had to be totally attentive to prevent a disgruntled bill from taking off a foot, and every step or so was rewarded with a pop and a squirt of sticky yolk up the trouser leg.

My constant companion while filming around Costa Rica was Julio Rivera, naturalist and television fixer extraordinaire. Julio is an amazing fellow, a mystical Latin American with a touch of magical realism about him. He left his family of 16 brothers and sisters at six years old in order to walk to the Caribbean and find out if a black man's skin was paint and smeared to the touch. His parents thought him dead, and when he returned three months later they celebrated his return by working him with the cattle for 14 hours a day. Midway through the day he would ride away to town and put himself through school by renting out his horse to the rich children. He got his first pair of shoes at age 15. As a young man he died after being bitten by the deadly fer-de-lance snake and was brought back to life three days later convinced he had seen God. He then didn't cut his hair or beard for a decade. He knows every plant, tree and creature in this, the most biodiverse country in the world, and his new wife has nicknamed him the Queen, as his travels are constantly interrupted by continual waving – everyone in the country seems to be a personal friend. He stands about three inches shorter than me but seems far taller, broad in every direction and with a large face, large nose and loads of hair. When he speaks English, it is with a charming, extravagant Spanish accent, and it seems there is not a single problem he can't fix.

Julio was responsible for getting me here at the perfect time, for possibly the largest *arribadas* of the year, with an estimated 75,000 turtles coming ashore in three days. Within a few square metres you could see every part of the process taking place. One turtle was digging with her broad foreflippers, another turned to scoop out a deep well with her rear flippers. Behind the next turtle, you could watch the mucus-covered ping pong balls plopping into the hole, then one turtle covering them, and another waddling her heavy carapace over the sand to flatten it. You could even see tiny matchbox-sized hatchlings flapping for the sea and trying to avoid being run over by the approaching juggernauts.

As the shadows lengthened, we avoided using lights on the beach so as not to distract the turtles. Instead, we used infra-red cameras, which made the whole event appear to be some primordial orgy, and made me look like a grinning, cat-eyed demon.

As it got darker and darker, the numbers just kept increasing. It got so that you couldn't take a single step without killing a turtle's potential progeny. Mind you, every egg you stepped on was a drop in the ocean. With so many turtles in one place, single nesting sites can get turned over as many as 30 times – that means one mother lays her eggs, then another comes in and digs them up while preparing her own nest (the eggs that are dug up all rot and perish). Even more dramatic is the predation. The beach is absolutely rife with life. Flocks of black-headed and turkey vultures a hundred birds strong sit behind the mothers and peck up the eggs as they drop. Dogs, raccoons, coyotes, night herons and roseate spoonbills dig up the nests, devouring everything. As the hatchlings emerge from the sands, they have to evade every scavenger from a 50-mile radius wanting to make them a light appetiser. The local villagers take half a million eggs every night of the *arribadas* and have a huge party on the beach, bagging and washing the leathery little ping pong balls for shipping off to restaurants all over the country. At least collections are well monitored and considered to be sustainable, which is more than can be said for many other countries around the world, where eggs and turtles are harvested remorselessly.

Less than one in a thousand hatchlings will actually make it through to maturity, meaning each female probably only manages to produce a handful of successful young in a lifetime. It's unimaginable that the expenditure of such energy can result in so little reproductive success, and also that after 220 million years on the planet, marine turtles that spend their whole life at sea can still be tied to the land for one part of their life cycle. However, the process continues still. Here on Ostional beach, while the predators sit too bloated to move, a few hatchlings succeed in that final scamper into the surf. Some are making their first frantic splashes on the road to majestic adulthood, and will be returning to these same sands to dig their own nests in a decade's time.

• • •

With essentially the same strategy for reproduction, the mortality rate of eggs from croc nests is also ludicrously high. Nests can be flooded, baked or excavated by a whole range of different scavengers and predators. Monitor lizards are so named because they are said to monitor the presence of crocodiles to find out where their nests are. Then when the young reptiles hatch, they have to battle to the surface and make it to the water without getting munched. Everything from gulls to herons to village dogs hang out for the easiest meal around. No single nest is guaranteed to have one hatchling make it to maturity.

This does however mean that there is one very simple thing that can be done to ease the plight of any struggling croc or turtle species. Simply excavating the eggs directly after hatching and then taking them to a hatchery where they can be incubated in safety improves survival prospects significantly. If the youngsters are then kept and fed until they are a year in age, their chances jump even more. I've spent countless hours with hatchery projects around the world, with one leading to the most ludicrous experience of all: hanging on a line beneath a flying helicopter over the Northern Territory before being dropped onto the top of saltwater crocodile nests ... with the females waiting there with jaws snapping!

Far more common nowadays though would be the turtle sanctuary I worked with in Ecuador, based on the edge of a marine reserve and focused not just on hatching successful juveniles but on rescuing beleaguered adults. In a beachside facility under thatched roofs were dozens of makeshift plastic ponds, each of which was home to two or three ocean giants that had seen better days. While some of them were struggling after boat strikes or from malnutrition, by far the majority had been tangled in plastic fishing gear. Many of them had had their flippers completely removed after a fishing line had sliced through them, leaving nothing more than stumps.

During my visit, the team and I were at a dive site inside the national park working with giant oceanic manta rays when a green turtle swam past. From its beak a thick monofilament line at least ten metres in length stretched off in the water. I instinctively grabbed hold of the line, trying to reel the animal back in towards me so I could get a hold of it and pull the hook out of its mouth. I was, however, totally unprepared. The turtle dived and was surprisingly strong, pulling me down with it. With both my hands full of line, I couldn't equalise my ears, which seemed bound to pop. The line sliced through my ungloved fingers … and I let go. The turtle swam off into the green depths, and then there was silence. None of the team said anything. We all knew the score. The turtle was doomed, condemned to trail that line from its mouth until it tangled around a flipper and probably killed it. I felt I had as good as consigned that beautiful and ancient creature to its doom.

Three days later we were diving about a mile away, at night and at a totally different dive site. It had been a sensational dive, with humpback whale song haunting through the water during the whole experience. We watched a moray eel hunting in the crevices of the reef and had giant pufferfish come and sit in our laps on the bottom. But then it got even better. Suddenly, a large green turtle swam past before turning and coming straight back to me, slowing down and dropping down to the bottom in front of me. Trailing from the side of its beak was ten metres of mono-filament line. It was the same turtle. This time it didn't swim away, but just lay there looking at me. Turtles just don't do that. It was as if it was inviting me to help. This time I was prepared. Grabbing the rusting hook in its beak, I manhandled the turtle's carapace and ripped with all my might. The line came out, the turtle thrashed away and was free. At first it swam straight upwards at speed, clearly hurt and shocked by the experience, but just minutes later it swam back down and leisurely past me. It would take a cynical soul not to feel as though it were trying to say thank you.

• • •

Two years ago I found myself walking down the paradise beach of an uninhabited Pacific island, musing that I was in a land that time had forgotten, countless miles from the nearest human being. The secret I did not reveal to the camera was that the crew and I had spent the previous hour using palm fronds and my soundman's boom mic as makeshift brooms, sweeping the strandline sands clear of soda bottles that'd bobbed over from the other side of the Pacific. We'd picked up sun-bleached Tonka truck toys from Spain and creepy glassy-eyed dolls, like something out of a jump-scare horror movie. The beach had looked like a toddler had just had a tantrum with their toy chest.

This modern flotsam and jetsam is borne around the world on the currents and maelstroms, and deposited on countless beaches in some infernal multi-coloured trash slick. In gyres the size of nations, forlorn plastics bob, turning our planet's biggest habitat into a technicolour graveyard.

Researchers studying 34 different species of seabird found that three-quarters had plastic gunk in their stomachs. A mighty male sperm whale washed ashore on the coast of Scotland a few years back and had a ball of plastic in its stomach that weighed more than 100 kilos. In the waters around the British Isles, leatherback turtles* suffer because they feed on jellyfish. With their bleary eyes, turtles are not known for their visual acuity (I once saw one trying to mate with a floating marker buoy), and a plastic bag swept on the currents looks much like jellyfish. The backwards-facing spines in their throat, designed to prevent jellyfish washing back out of their gullets, trap the plastic in place, and they suffocate.

---

* Yes, there are marine turtles in the waters around the British Isles. I've had a leatherback paddle into my kayak in Cornwall, and the largest ever was found in Ireland – it weighed more than a tonne.

I proposed to Helen in the Southern Africa nation of Namibia, but one of our most powerful memories of the trip was spending a day chasing around a young fur seal we'd encountered while out sea kayaking. It was wrapped around in packing tape that had clearly been cutting through its growing flesh for many months and was slicing it down to the bone. We finally managed to pin it down, drag it ashore and cut the plastic free, but not without receiving a few lacerations – the little blighter wasn't nearly grateful enough. Anyone who doubts packing tape can cut through growing flesh should try opening an Amazon package without a pair of scissors. Or imagine ripping apart the plastic rings that hold together a six pack of beers without any hands. These can holders are the nemesis of every wild animal, terrestrial and aquatic, so please shred them before discarding them.

Closer to home, Helen and I have teamed up with a local wildlife refuge and now release rehabilitated wildlife in our back garden. A good portion of the animals we take on will have either eaten or been caught up in plastic detritus. And how does this link up to marine plastics? Well, 80 per cent of marine litter originates on land. There are a multitude of ways that can happen, from windblown bags on landfill to good old-fashioned litter. A soft drinks bottle dumped in a Scunthorpe street can easily be swept into a storm drain, then into a river and out to sea, before slowly making its way around the world till it chokes a dolphin in Australia. If that seems far-fetched, consider that a plastic bottle takes at least 450 years to break down, and individual pieces of plastic have been traced trundling their way round the world, until they finally get stuck in ocean gyres that can stretch for hundreds of miles.

So what can we all do to make things better? Well, the chances are you've probably already done something; for example, taking along your own bag to a supermarket instead of paying for a single-use one. The plastic bag tax has been one of the single biggest triumphs in conservation of my lifetime, taking billions – and, yes, I do mean billions – of single-use plastic bags out of the ecosystem every year. Every time you forget your bag for life, think of a hedgehog wearing your discarded bag as a hat and don't forget again!

A staggering 91 per cent of plastic is not recycled,* and that's a situation we have to rectify before we end up drowning under mountains of soda bottles. Litter doesn't just make our country ugly, it kills wild animals, and not just here, but on the other side of the world too. It's lazy, it's ignorant, it's destructive, it's inexcusable.

You could go one step further and campaign to take your local shop, or even your local village, plastic bag free. Paper bags are perfectly good for almost everything we buy nowadays, and the more we can wean ourselves off single-use bags the better. And if your favourite coffee shop or greasy spoon uses single-use stirrers and sachets, encourage them to start using refills and cutlery; it's the future, as well as the past.

Until the government ban on microbeads in cosmetic products comes into effect, do not buy and do not use any product that contains plastic microbeads, and don't purchase them in other countries. Doing so means flushing billions of microscopic bits of chemicals out to sea and into the fish we eat. How these monstrosities were ever legal is beyond me. And now these fripperies of modernity are colliding with beasts that have been around since the genesis of invertebrate life.

Recently I woke at dawn on an Indian Ocean island, my family nuzzled alongside me. Going out onto the beach, I saw

---

* This high statistic is caused by both improper disposal of plastics by consumers and the fact that 25 per cent of the plastics we produce are thermoset, meaning they do not soften when exposed to heat, so they can't be melted and repurposed.

tank-like tracks running up the sand and sprinted back to get the kids. For them, watching that green turtle cover her nest and return to the waves was awe-inspiring – a moment that cast them back in time and might conceivably be their first long-term memory. And after we returned to our room, I took myself aside and wept. Every encounter with a marine turtle leaves you humbled. There are a thousand reasons for protecting the future of these ancient mariners, but here I felt it most keenly, in the unbridled wonder of my youngsters.

CHAPTER NINE

Spineless Wonders

A lot of people say an octopus is like an alien.
But the strange thing is, as you get closer to them,
you realise that we're very similar in a lot of ways.

Craig Foster, *My Octopus Teacher*

I was in the Far East and I went into a restaurant and I
ordered octopus and the waiter said: 'It takes four hours.'
I asked why and he said: 'It keeps turning off the gas.'

Frank Carson

*Inky blackness and torchlight, reflected by the water, spins the senses.
Panicked voices on the intercom. Beyond the cut of our torchlight is
the unknown. Grotesque, demonic monstrosities twist in my fever-
dream brain. Normally I would take a deep breath and remind
myself to be rational and calm, but we are here for an encounter*

with a real-life sea beast. The Kraken himself. We know that somewhere below the surface is a fantastical kicking form: El Diablo Rojo, the red devil, commonly known as the Humboldt squid.

Time is ticking. We need to be in the water now. Final checks to the equipment, a layer of baby shampoo over the glass of our masks to prevent fogging, sound checks done. An industrial harness is pulled over a $10,000 chain-mail suit, snagging on the links, threatening to cost me a fortune in repairs. 'Don't let them rush you,' I keep saying to myself. 'Sloppiness kills more divers than sharks ever could.'

'Time's running out, Steve. It's now or never!'

I struggle to stand, with my huge, heavy twinset of dive cylinders and medieval armour. Strong hands grip my upper arms and steer me towards the dive deck. I shuffle forward, clumsy in my fins, like a swan waddling along the waterfront.

'Three, two, one, dive!' and I'm in. There is a splash of white water, and static rages in my ears from my communication system. The lights inside my mask are too strong, and all I can see is my own reflection on the inside of the glass. My only sense of which way is up and which is down comes from the wire hawser linking me to the boat and safety. This might seem like overkill, but the last time cameraman Simon filmed these alien assassins, he was grabbed by the ankle and pulled down into the depths, bursting both his eardrums. He only managed to escape with a few stabs of a hastily brandished dive knife.

I fight to control my breathing and recover my sense of equilibrium. Dive guide Scott is urgently calling me over.

'Quick, Steve, I have one! A bunch of others were up here hammering it!'

Humboldt squid are known for cannibalism, and many that are brought to the surface have been devoured alive by the very compatriots they had minutes earlier been hunting alongside. Then I see it, held by the mantle between Scott's hands. It far surpasses *its reputation – it is as big as a person, arms and feeding tentacles pulsing wildly in the water. The entire body is flashing from white to bright red, over and over again. Scott beckons me closer, and transfers the Kraken into my chain-mail gloves.*

• • •

There are few animal groups that have enjoyed such a renaissance in public perception as the cephalopods. Roughly translated as 'head-footed', these are the octopuses, squids, cuttlefish and nautiluses. Individual octopuses have become major attractions at aquariums around the world, with obvious defined personalities and some able to perform Houdini-like escape feats and stunning tricks. These weird alien creatures seem like the most different beasts on earth to us. Some grow from an egg the size of a grain of rice, to the size of an adult human, and then die all within three years. However, they have taught us vast amounts about our own neuroscience, the ageing process, evolution … truly the octopus is our teacher. It's no coincidence that when Jurassic Park wanted to make their *Indominus rex* powerful beyond all measure, they gave it the camouflaging DNA of a cuttlefish.

While Helen was away competing at the Tokyo Olympics in 2021, my good friends at the National Marine Aquarium in Plymouth allowed me to take our three kids behind the scenes. We met their giant Pacific octopus, who clearly knew his keeper, coming across the tank to greet him. It turned out that the aquarium had to provide enrichment for their octopuses, much like they might a captive monkey; if they didn't, the animals would get bored and cause mischief. The octopus then proceeded to entertain the kids by juggling with a revolving toy while keeping it in motion, before opening a closed jam jar to retrieve a tasty shrimp treat from inside. These tricks had not been taught but picked up through trial and error. Once the octopus had made one mistake, it never made it again, but learned through experience.

I was blown away. As our twins Kit and Bo were only one, and Logan three, they just assumed that all squidgy invertebrates behaved this way, which they don't! Cephalopods, though, break the mould. There was an octopus at a New Zealand aquarium who learned to take people's photographs and another in Seattle who used to blast one female keeper with icy water every time she came in to feed it. For some reason, it decided it just didn't like her.

The evolution of cephalopods is one of the best documented of any sea creature, with the drawers of many institutions of learning filled with 500-million-year-old Cambrian fossils of extinct cephalopods that bore shells. These are some of the most prevalent and easily discovered of all fossils. There are well-used seacliff climbs down on the Dorset coast near Dancing Ledge, where the belayer can stand with their damp feet on top of fossilised ammonites the size of dustbin lids. There are several Norfolk beach walks where you can come back with pockets full of what look like stone bullets – actually the fossilised shells of belemnites. These two groups (the belemnoids and ammonoids) were joined by the nautiloids, which are still represented in our seas today, free-swimming in the water column with their intricate and beautiful shells. Because our best records are made up of these shelled sea beasts, this might actually skew our understanding towards the animals that had hard – readily fossilising – body parts rather than those that did not.

Our present knowledge suggests that the 700 species of cephalopods we see today emerged in the Mesozoic period, 160–100 million years ago. Fish, squid and the predators that fed on them were all locked in an evolutionary arms race at this time, getting faster and developing more lethal weapons and tools, such as camouflage and ink secretion. It seems this was when the majority of cephalopod species lost their internal shells or skeletons, in order to become more agile, and more able to cram themselves into tiny spaces to escape their predators.

Though octopuses are the glory hounds stealing all the publicity, squid are the most numerous and successful of modern cephalopods, with some shoals of firefly squid numbering in the hundreds of thousands, and opalescent inshore squid coming together in their millions during breeding congregations. They range in size from the tiny but entrancing Thai bobtail squid, whose male is the size of my little fingernail, up to the mighty giant and colossal squids that can be double the length of a bus. They can be found everywhere, including the deepest oceans, where Magnapinnidae have been seen at depths of 5,000 metres plus.

At those depths, they live long and slow, and may have weird lives – and they don't come much weirder than *Vampyroteuthis infernalis*, the infernal vampire squid. It has the largest eyes in the animal kingdom in proportion to its body size. They are usually blue but can transform into burning red, justifying the 'infernal' moniker. When threatened, they will bite off their own arms, which float away glowing with bioluminescence to distract a predator. When they eject an ink cloud, it too is filled with bioluminescent particles, creating a veritable shower of light.

Other species practise diel vertical migration and can be found up to the surface, where many species will aggregate, especially at night. Squid are found in multitudes in the tropics, but can also be found in polar waters, where species like the southern glass squid thrive. This is also where the mighty colossal squid was first discovered, with one being brought to the surface alive in 2007.

For all their success, they are also one of the most hapless victims and valuable food sources of our seas. Pretty much every aquatic predator will feast on these packets of protein at some period in its life cycle, including other squid. Mighty ocean giants such as 60-tonne sperm whales and 4-tonne elephant seals both have a diet that is 95 per cent squid; elephant seals around South Georgia consume 2.2 million tonnes of the stuff every year!

The majority of cephalopods don't tend to survive long, with most squid species dying before they reach a year, and few living longer than three. They are semelparous, having only one mating event before senescence and death. The male transfers a sperm packet to the female using a specially developed arm called a hectocotylus. In some species this is fired directly into the mantle of the female; in others it can be fired into cavities around the eyes. There have even been reports over the years of these sperm packets firing into the mouths of unwary diners. One peer-reviewed scientific paper describes a 63-year-old Korean woman who was eating poorly cooked squid and suffered immediate and severe pain in her mouth. She was rushed to hospital and operated on, and was found to have 12 'white spindle-shaped, bug-like organisms stuck in the mucous membrane' of her tongue and cheek.

· · ·

Perhaps my best squid learning experiences have been in the company of American biologist and eccentric Scott Cassell. White-haired and charming, Scott has a barrel chest and knotted biceps that give him away as ex-military, even through his street clothes or wetsuit. A former counter-terrorism operative, army sniper, medic, biologist, explorer and inventor, Scott is a fron-tiersman character, much like the pioneers who drove forward the genesis of diving. He has built a suite of his own submarines (which he has been trying and failing to convince me to use for over a decade) from backyard scrap. He has also done more dives with Humboldt squid than anyone else alive.

In the labs of the local university in Baja, Mexico, Scott and I were given the opportunity to do a dissection on a fresh Humboldt to get a sense of why they are so awesome. He began by taking his scalpel and slicing down the length of the mantle of a Humboldt squid to expose the mantle cavity, a space which contains all the important stuff: gills, viscera, gut and kidneys. Below the mantle sits the funnel or siphon, a tube which appears to be (with the arms and tentacles) an adaptation of the molluscan foot. This siphon is used to power water in order to blast themselves along at high speeds. It can also be used to clear sand from the bottom when hunting, and the females use it to blow clear, aerated water over their eggs when they're guarding them.

'Cephalopods are molluscs,' Scott said, 'like snails, and bivalves such as clams.' It was hard to reconcile this simple fact with the animal I'd seen zipping around me, solving tasks and mazes, and even with the alien being dissected on the slab in front of me now.

'Obviously they're much more advanced than most, but there are many things they have in common. And this is one of them.' Scott pulled the arms aside and uncovered the beak, within which was the radula, a toothed ribbon that moves backwards and forwards like a rasp. All molluscs have one. It is usually a tongue-like organ covered in velcro-like teeth which can be used to grind algae off rocks, although in some (like the highly venomous cone shell) their salivary papillae have evolved into venom-injecting lances that can be used to stab their prey, and in the netted dog whelk, it is used as a drill to bore clear through the shell of other mollusks. In the squid in front of us, the radula was surrounded by a beak that was highly flexible and light brown at its base, and hard as iron and black in colour at the tip. It's a formidable tool. Scott showed it opening and closing. It was a bit like one of those paper fortune tellers that children make.

'So it squeezes the beak using a sphincter muscle around the outside, which is really efficient. That creates some slicing power here ...' Scott emphasised the scalpel-sharp cutting edge of the beak, 'but a quite phenomenal force at the tip here.' He was right. Even gently squeezing the beak in my hand, I could generate enough force to snip off a finger with the tip.

This weapon is supplemented in many species with virulent venoms. In the common octopus it's a toxin that can break down the shells of their crab victims, complemented with a 'chitinase', an enzyme that separates the shells from the muscles. The most virulent, though, is surely Australasia's blue-ringed octopus. No larger than golf balls, they are extremely tricky to find. In fact my most successful hunts have been on the seabed near Australian pubs, where post-closing time boozers have taken to hurling their beer bottles into the sea. The tiny octopuses curl up inside the bottles, taking shelter and solace, and only coming out to hunt. When you first find them, they usually appear tiny, innocuous and probably dull brown in colour, matching the tinted glass of their beer bottle homes. However, the second they are exposed, they transform into a bright yellow and seem to double in size. The yellow is a backdrop for their brilliant aposematic display: black and neon blue circles that pulsate on the skin and warn the unwise of the lethal toxins they possess.

These little beauties use tetrodotoxin as a venom. It is one of the most virulent toxins in nature, and can also be found in the skin and organs of the pufferfish. Japanese diners famously consume this as fugu. If prepared correctly, with just the right amount of tetrodotoxin present, the eater will get a faint tingling on the lips and tongue, known as the 'brush with death'. A handful of diners who don't consume fugu with the right mix die every year.

Aquarists describe trying to keep blue-ringed octopuses in captivity and coming down the next morning to find everything else in the aquarium floating motionless on the surface. As part of my academic studies into venomics and toxicology, I've been through all the data relating to marine animal envenomation and found very few examples of a blue-ring ever harming a person. One of those unlucky few was a female swimmer who found one in a rock pool, posed for a photo with it in the cleavage of her bikini and never even felt the bite. Which is a pretty good case for a Darwin Award.

In the cartilage just below the brain, Scott unravelled some complex paired organs. 'These are the statocysts,' he told me. 'They're kind of like mechano-receptors. If you go in and mess with these little things in a living animal, it just kind of swims around in circles bumping into stuff. They tell the animal information about which way is up and down, about gravity, acceleration and things, helping them compensate movements in the head and eyes. It's also probable that these have some kind of function in regulating their colour – so creating countershading.'*

'So the animal can see the colours in the environment around it, process that information and then change colour to match?' I asked.

'Well, that's a weird one,' Scott mused, 'cos most research seems to show cephalopods are colour blind. They have no rods or cones in their eyes, and only two colour pigments. Mind you, they do have photosensors all over the body, so it might be more down to those.

'OK, pop quiz,' Scott continued. 'Number one fact people know about octopuses?'

I pondered. The camera was rolling. I didn't want to make a fool of myself. 'Well … I guess that they have three hearts?'

---

* As briefly mentioned in Chapter Three, countershading is a dark topside and light underneath to match these shades in the seascape.

'Correct! A main central one, and one by each of their two paired gills.' Scott teased the gills out of the jelly mess on the table. They looked like a mille-feuille pastry that had been over-glazed in gelatin.

'So they pump oxygen from water into the gills, and need to move that oxygenated blood through their bodies. Instead of our red haemoglobin, which is red cos of the iron, they have haemocyanin, which contains copper.'

'So they have blue blood?' I asked.

'Yes, and that's just the start of the weirdness. We have just one brain, right, up here in our skulls.' He knocked on his head as if to check his was present. 'But lots of invertebrates have ganglions of neurons that serve the same purpose, but may be spread out through the body.'

I nodded. Around 60 per cent of an octopus's brain is located in its arms, enabling it to respond to stimuli much faster and independently of the 'main' brain.

'It's the most amazing thing,' Scott continued as he probed around in the milky jelly. 'The nervous system of the arms contains more neurons than the rest of the nervous system put together, and if an arm gets cut off, it keeps moving for hours afterwards!'

This is not uncommon in nature; for example, many reptiles drop their tails in response to a predator. However, while a lizard's disconnected tail appears to be twitching from nerve programming to distract a predator, a sliced octopus arm appears to be acting, even thinking, for itself. They may even be able catch hold of food, though detached from the mouth that could eat it!

'Ah! Here we go!' Scott was triumphant, as if he had just struck gold. 'Can you see that? You are looking at one of the great miracles of the natural world, my friend!'

I peered down. His blue gloved hand was tickling what looked like a piece of dental floss, stretching from the top of the mantle down. It looked deeply unimpressive.

'It's a giant axon!' he declared, as if this was the most obvious thing in the world. 'Like a nerve on steroids!'

Giant axons are one of the most intriguing aspects of squid anatomy. These nerve cells, which can be as much as one millimetre in diameter, are radically thicker than those of other animals. Scott teased one out of its surrounding tissue. They were first described in the 1930s by anatomist John Zachary Young, and connect to giant cells in the brain. Much like a fibre-optic communications cable, bigger axons mean faster traffic – that is, faster signals being sent around the squid's body. The best thing about these giant axons is that their size has allowed us to study them, and therefore learn so much of what we know about nerve cells, even those in human anatomy.

• • •

The next day, Scott and I headed out into the Sea of Cortez from the Port of Ensenada, itself an intimidating place, beset by a crystal meth epidemic that makes it a place no outsider wants to tarry long. At sea the temperature was easily the hottest I'd ever experienced – it was windless and hovering around 40°C, which is utterly remorseless when you're on deck in a neoprene wetsuit. Over the top of the rubber, we were clad in full-body chain mail suits, harnesses and steel tethers connecting us to cranes out on the boat so we could be hoisted out if anything went wrong, all of which only intensified the heat. 'I've hand fed great whites and bull sharks,' Scott said. 'But the Humboldt squid is the most dangerous being I have encountered … well, other than maybe a lawyer.' I laughed, though I sensed he might have used this line once or twice before.

'This beak is where all the damage is done,' he said, holding his hand pinched as if inside a sock puppet. 'It's kind of like a giant parrot, but one the size and weight of a large person. Imagine being bitten by a parrot the size of me! I've seen this go through bone in a tuna's skull.' This was a combination of images I couldn't begin to put together.

I knew Scott well enough to ask him questions that others might have considered disrespectful. 'There are lots of biologists,' I picked my words carefully, 'who study Humboldts, and have dived with them in T-shirts and shorts. They think their danger to us is overstated, that all this kit', I motioned at the chain mail and harnesses, 'is just for show and totally unnecessary.'

Scott smiled. He wasn't offended – far from it. It gave him the chance to use his trump card. 'I've heard those guys too. Most of the time they're just diving with juvenile Humboldts, which is kind of like playing with a tiger cub then claiming that tigers are harmless. The small ones are never going to go for you.

'I've done more than 2,000 dives with them. Sometimes they'll just swim by you and barely look at you, and other times they'll attack you with a ferocity that is unrivalled in the natural kingdom. Who knows why? This is not a safe animal to be around. Maybe it's water temperature, when they're hungry, season, whatever – we know so little about their lives. But get a Humboldt squid when it's in predatory mode, and it's the scariest thing you'll ever see ...'

Scott then leaned forward and parted the hair on his scalp, showing off a wicked scar below his white hair. I felt as if I had just become an extra in the scar comparison scene in *Jaws*.

'This big sucker came at me from above – I never even saw it coming. The bite cut through my dive hood, and right down to the bone, so I could feel cold seawater on my skull. The pain was insane. I tried to wrestle it off me, but it has suction cups all down the arms, ringed with chitinous teeth.'

He paused. Chitin is a common protein in nature – found in insect exoskeletons, for example – capable of being super flexible or as hard as metal if it is tanned or sclerotised. The ring teeth in the sucker cups of a Humboldt squid look like a piranha's.

'An adult male could have 1,200 sucker cups, each sucker cup has 36 teeth, which means one squid could be using 36,000 of them to hang onto its prey.'

'So once it's got ahold of something there's no chance it's getting away,' I said.

'Correct. It has one of the best grips in the whole animal kingdom. First off I tried to throw the thing off me, but it grabbed my wrist and broke it in five places. Then it got hold of my throat and ripped it open. I superglued my scalp and throat back up …'

I was wide-eyed and incredulous, but he still had the punch-line to finish. 'But I had to go into hospital to get the wrist reset. Waste of time that was.'

It would be easy to ignore Scott's stories as bravado. After all, he is a luminary of Discovery Shark Week and has been a consultant on such schlock-fests as *Sharknado*. His legend is his livelihood. However, on our second expedition filming Humboldts together, Scott proved to me why these beasts should be both feared and revered.

It was one o'clock in the morning, and we'd been fishing off the back of the boat for four or five hours with no success. Three hundred metres of line descending down into the depths, with a dozen wicked-looking luminous lures attached, had to be constantly jigged up and down to attract any potential customers to strike. Then, finally, we got a bite – the line was wrenched out of our hands, and we had to haul for ten solid minutes to get the Humboldt to the surface. Once we had landed our sea monster, it was all hands to the wheel to complete our mission. Scott had rigged a sort of specialised squid harness, with a live lipstick camera attached that enabled us to see what the animal saw as it descended back down to its inky home. This had to be attached quickly, with the animal still in the water to keep it cool and its gills aerated. As soon as it was complete, Scott shouted 'Clear!' and we released it.

With a super-soaker squirt of water from its siphon, the squid was gone, dropping from sight. Down, down, down it went, us paying out line as it swam deeper, 200 metres, 300, 400, into a realm where we could never follow. We'd rigged the camera with a red light to illuminate the way, and it picked up all the plankton in the water. With the spaceship-like mantle of the squid in the foreground, these looked like stars shooting past a sci-fi vessel travelling at warp speed. And then it levelled out, and out of the darkness shapes started to loom and coalesce into clarity. In perfect symmetry, equally spaced apart, 'flying' in formation like a squadron of Second World War bombers, were Humboldt squid. Our subject fell in and joined the ominous platoon. There could have been hundreds of them. If any single one of these was capable of shredding a beast of a man like Scott, a fighting force like this must surely be one of the most indomitable on earth.

Underwater, the Humboldt is a fearsome beast. The ones we filmed had been drawn up into the shallows by lures or were cannibalistic opportunists who were following the others to take sneaky bites out of them. They're big – the largest I've ever seen up close was maybe seven feet in length, but they can grow to more like nine. Their arms are short – certainly if you compare them to their giant squid cousins – and in the centre of each is a circular maw with waving white tendrils around it. This houses the beak, which as Scott said really does look like a giant parrot's beak.

This was a part of the squid's anatomy I was already acquainted with, as I'd had a close scrape two nights before. That was my first dive with Humboldts. As Scott had transferred the giant animal to me to affix the camera harness, it had twisted and sunk its beak into my arm. The bite had been agony, even through the expensive chain-mail suit, tearing two holes in it, and I could feel that if it had been a little lower down and on the bones of my wrist, they'd have been splintered like matchsticks. As I'd fought to control the thrashing beast in my hands, I'd been aware of pulsing white shapes below me – the other squid lunging up towards us, as if building up the courage for a strike. It had been eerie beyond words.

Back in 2008, when I had this experience, the Sea of Cortez squid fishery was at its height. Fifteen hundred boats focused purely on catching the red devils and countless tonnes caught throughout the season. We waited no more than minutes in our bobbing wooden panga for the local fishermen to catch one and tow it to the surface, and their efforts were constant, dropping the jigs down and pulling them back up with squid attached. By 2011, when I came back to film the animals for a second time, the squid were just as big, but it took an awful lot longer to find and catch one. In 2015, when I was thinking about returning once again to film, the message from Scott was clear. 'Don't bother. The squid are gone. Jeremy Wade from *River Monsters* came out here and didn't get anything bigger than a shoebox.'

The fishery had completely collapsed, and has not recovered. The reasons are not clear. Some have blamed climate change and changing ocean currents. Others say it is because of exploitation of squids' prey. Either way, catching hundreds of tonnes of an apex predator without really understanding much about its life history is bound to lead to failure. It's particularly sad when you learn that most Humboldt squid is not fit for human consumption and ends up being mulched as fertiliser for agriculture. This undignified end seems all the more wasteful when you properly start to unravel the miracle of this improbable predator.

• • •

The final element to our character study of the Humboldt squid was working with PhD researcher Hannah Rosen,* who was researching their use of colour change for communication. She set up a makeshift lab inside our dive boat and took some fresh slices of squid mantle from out of the fridge. Stretching them out to their full extent, she then took two electrodes and attached them to either side of the skin, passing current right through it. The result was utterly extraordinary, like seeing ink dripped into a glass of water and blooming into clouds of colour.

The colour spread instantaneously across the sample, seemingly as a uniform wave of pigment. However, when you looked closely you could see the skin was covered with tiny dots. When stimulated with the current, these dots dilated or flattened out, making more or less of their colour visible. These were pigment-containing sacs called chromatophores, and they work in tandem with light-emitting cells called iridophores.

'When the squid is dark red you wouldn't see it,' Hannah said, 'but then it flashes white and stands out – it's the opposite of camouflage, so it has to be some kind of communication, right? We see this when they're hunting and when they're mating, but unfortunately we don't know much about why. They just live too deep for us to figure it out.'

The cephalopod that uses colour change to greatest effect is surely the tiny cuttlefish found on sand flats in the fast-flowing straits of Indonesia. The aptly named flamboyant cuttlefish is no

---

* Now Dr Hannah Rosen, a world-renowned expert on the communication of cephalopods.

bigger than your thumb, but it is one of the most intriguing, charming and impossible of all marine predators. To find one, my film crew headed to the islands of the Komodo National Park, but rather than focusing on the legendary dragons that live here and nowhere else, or on the manta rays that so often buzzed us inquisitively as we searched the seabed, we swam away from the beautiful coral gardens and out to the apparently barren mud flats, scouring in among small clumps of weed and rock piles by day and by night.

On the third day of searching and finding nothing, we were forced to dive in current I would never normally consider. The straits between these islands rip at phenomenal speeds, causing whirlpools and rampaging tide races to form – it has to be taken very seriously indeed. In fact, the only place we could even consider diving was in the lee of a large rocky outcrop, about the size of a house. All around us, the flow was so intense that it would have ripped the mask clean off my face. Every one of us wore locator beacons and carried surface marker buoys, to prepare for someone getting separated and swept off into the South China Sea.

Towards the end of the dive, cameraman Simon called out in excitement, 'Steve! I've got one!' I swam along close to the sandy bottom, anchored myself with a hand into the silt and looked where Simon was pointing. And looked some more. I couldn't see anything. It was only when the flamboyant moved that I could actually see what he meant. Not only was it very small – half the size of a matchbox – but before it moved, it was the exact dull green and brown colour of the seagrass stump it was pretending to be. To finish off its disguise, the flamboyant's whole body was contorted into the shape of its environment, arms and feeding tentacles lifted above its eyes into seagrass tendrils, mantle puckered into textured folds.

Once out in the open, the transformation was both bold and instantaneous. Now the base colour was a luminous white, but waves of indigo stripe passed down the length of its body. To begin with, it was a little like rippled sunlight along a sandy seabed, but then the colours became more intense, more hypnotic. Two clubbed antenna came into view at the front of the animal and it tensed, poised like a boxer waiting to throw a hammer blow. Then the two feeding tentacles fired out simultaneously, capturing a tiny shrimp from a body-length in front of the cuttlefish. It repeated the process over and over, with 100 per cent success every time. These feeding tentacles have been measured to fire out in a hundredth of a second.

Suddenly the cuttlefish began to move along the bottom, displaying a kind of locomotion completely unique to this species. Rather than using its siphon to blast water, or moving the fins down the side of its mantle, this thing galloped along the bottom like a rocking horse brought to life. It was quite the most extraordinary thing I've seen on the seabed. When it finally settled, the light display down its mantle took on a vibrancy and frequency that it is almost impossible to put into words.

I've seen this exact same display used on a different scale in the mating rituals of giant cuttlefish. At the right time of year, just off a beach in South Australia, these welly boot-sized cuttlefish come together in huge numbers to breed. The males intimidate each other and attract the females by turning onto their sides and bouncing these same dazzling purple ripples down their flanks.

As if they didn't have things hard enough, trying to both disco dance and street fight as the waves crashed them about on the rocks, on one occasion we took down a big mirror and used it to show the males their own reflections. They immediately faced off against themselves, turning broadside so as to display as much of their light show to their rivals as possible. They would furiously and angrily demonstrate their firework-like expertise before finally butting into the glass to try to drive their own reflections away.

．．．

While all cuttlefish species – even the so-called giant – are comparatively small, that is not true of all cephalopods. At the other end of the scale is one of the most iconic of all sea beasts, but one of the least known and understood. It's one of the few animals I talk and write about regularly while knowing that I will never see it with my own eyes. The label giant squid usually refers to animals in the genus *Architeuthis*, from the Greek *archi*, which means 'chief', and *teuthis*, meaning 'squid'. Biologists are uncertain how many species there are; some suggest there could be as many as 18 different kinds.

The most remarkable thing about these animals is how rarely they've been seen alive. Almost all of our records of giant squid come from specimens that have washed up freshly dead on beaches. Although the first record of a 'sea monk' washing ashore was in Norway in 1546, it was 1861 before the first acknowledged and accredited account of one swimming alive at the surface, an encounter that probably inspired Jules Verne's 1870 novel *Twenty Thousand Leagues Under the Sea*. The first good record of a physical specimen was in 1873, when Theophilus Picot and his crew rowed out in Newfoundland's chilly seas in search of herring. They ran into a vast gelatinous mass, with feeding tentacles that lashed onto their boat, entwining them and threatening to drag them under. The sailors hacked off one of these tentacles with an axe, leading to them all getting hosed down with gallons of ink. Miraculously, the animal swam away and seemed to survive. That tentacle, which was 5.8 metres in length, was preserved by the local rector, who was a keen naturalist.

The largest giant squid ever reliably recorded was the Thimble Tickle specimen, found stranded alive on a Newfoundland shore just five years later. Seventeen metres in length, it died as the tide receded. The fisherman who found and measured it cut it up for dog food.

In later years, whalers studied the sucker cup marks on the bodies of sperm whales and marvelled over the beaks they found amassed in their dead stomachs. They talked of the Kraken and the Great Calamari and told tall tales of seamen being dragged to their deaths.

Almost all records come from dead specimens, which have in extremis had mantle lengths of five metres. Extending the longest feeding tentacles to their full extent gives an overall body length of 18 metres. So if you laid one out on a tennis court, with the top of the mantle at the server's feet on the baseline, the edge of the mantle would overlap the service line, the arms would stretch to the net, and the feeding tentacles would go all the way over the other side and stretch out towards the person returning serve.

They weigh perhaps half a metric tonne and are the largest invertebrates ever known to have lived. In absolute terms they have the largest eye in the natural world, at approximately 40 centimetres in diameter, the size of a basketball. If they were just a scaled-up version of a Humboldt squid, then surely they would be the most terrifying predator in our seas. Sadly, science has rather poured oil on that hype. Their fins are not especially large or powerful, and the 'giant' nerve axons are smaller than those of the comparatively tiny common squid. Studying their statocysts has led biologists to infer that they hang motionless and neutrally buoyant in the water 'head down', waiting for prey rather than actively hunting it. They do have well-developed brains, beaks and arms, but they don't have hooks in their sucker cups like colossal squid do.

Giant squid do have ink sacs like many shallow water species. This ink is a mix of mucus and pure melanin, and can be squirted out when the animal feels threatened. The first, most obvious purpose for this is as a kind of smoke grenade, providing a screen behind which the squid can make good its escape. However, it might also function as a pseudomorph, or fake body. The sticky mucus causes the ink cloud blossoms to bind together underwater into one phantom entity, suggesting the appearance of an actual animal – either a giant foe that might scare off a predator, or an alternative prey that the predator might be distracted by while the squid makes good its escape.

Squid ink also contains dopamine – the chemical we associate with good times and relaxation. With predators such as sharks able to detect so many chemical processes through their olfactory systems, one theory is that this chemical might paralyse a predator's sense of smell. Or perhaps it is a pacifier, getting a predator so high that they forget to chase their victim? It's a wild idea …

It wasn't until July 2012 that the first recorded images of living giant squid – the most sought-after footage in natural history – were broadcast around the world. In grainy footage from a remotely operated vehicle deep beneath the surface, the giant squid was seen with its arms and tentacles lashing over a lure, the very image of an alien sea beast.

This surely is why the giant squid is still such a powerful presence in our imagination. All good horror directors know that if you want the greatest impact from your monster, you should show it as little as possible. Tease a little, and let our fevered imaginations colour in the gaps. The idea of a leviathan lurking in the depths, about which we know very little … it's one of the most powerful and exciting ideas in natural history.

CHAPTER TEN

Coral Symphony

Ice ages have come and gone.
Coral reefs have persisted.

Sylvia Earle

In the course of time I have learned to tramp about
coral reefs, twenty to thirty feet underwater, so
unconcernedly that I can pay attention to particular
definite things. But after all my silly fears have been
allayed, even now, with eyes overflowing with surfeit of
colour, I am still almost inarticulate. We need a whole
new vocabulary, new adjectives, adequately to describe
the designs and colours of under sea.

William Beebe

Our boat bounces over deep, deep blue velvet, the flattest seas you'll ever witness. We are in the Celtic Deep off West Wales. A dozen common dolphins bounce and roll at our bow, surfing our wave with grace. Fulmars and herring gulls ride the invisible pressure wave in the air above us. For a fleeting second, my eye catches something distinctive. A purple plastic bag at the surface? No ... it can't be. 'Stop!' I shout, and the captain powers down and swings the boat around.

On any other day we'd never have found the animal again. But today, with the waters like a vast sink of dark ink, we instantly spy it – a see-through crimson and cranberry Cornish pasty floating on the surface. The first Portuguese man o' war I've seen in our seas.

There's no hurry to get on our wetsuits – man o' wars are carried around passively by the wind, and there is none. Despite not needing to rush or stress, I can feel the blood thudding in my throat. This, after all, is an animal that could kill me or cameraman Mark with a blast from its stinging cells. We need to cover every inch of our skin so that no organic material can touch the cells and trigger them to fire.

We have a quick debrief – what we call a 'dynamic risk assessment' – chatting through our plan of attack. Then we slip into the water and swim over cautiously. The tentacles can stretch out in the water for ten metres, and even with neoprene between us and the stingers, we don't want to get entangled.

Up close, nervousness gives way to wonder. This is a miracle of nature. Despite its appearance, this isn't a jellyfish. It isn't even one single animal, but a colonial organism made up of combined units, each of which hatched from a single egg and then came together to form a functioning whole. I risk putting my mask-clad face into the water so I can look down at the killing threads below. Most of the tentacles are concertinaed up like dark blue slinkies towards the floating bell. Two dead fish have been ensnared and are being drawn towards its digestive system. However, there are also other

smaller fish swimming lazily within the lethal snares. These man o'
war fish live here, protected from predators by the virulent venom of
a floating fortress.

. . .

When many of us use the word 'animal', we do so with a certain
amount of mammalian bias, and the first thing that springs to
mind is probably a hairy, warm-blooded creature. But any zoolo-
gist will tell you that mammals make up a minuscule and irrelevant
part of global fauna,* numbering only a few thousand species.
Beetle species, on the other hand, number at least 80,000, and
insects as a whole might number 10 million different species.

You might also think that a prerequisite for an animal is that
it should move. Well, potentially the most important animal in
our seas spends most of its life barely moving at all – in fact, it
sits in a rock skeleton of its own making.

The corals are part of a group of around 11,000 aquatic
invertebrates in the phylum Cnidaria (with a silent C), along
with fellow lazy livers (i.e., not moving) the sea anemones and
sea pens. In the same taxonomic melting pot are jellyfish, box
jellies and colonial organisms such as the Portuguese man o' war.

Corals may have evolved 535 million years ago (though most
of those early species went extinct in the Permian extinction
event), and jellyfish at least 635 million years ago. Anything that
stays around that long on our planet is doing something right.
But much of the success of this group is not just down to adapt-
ability within species but to a sublime symbiosis with innocuous
little organisms called zooxanthellae.[†]

---

* Any ecologist or entomologist will tell you that our ecosystems would
do just fine without mammals, but take away the bugs and there'd be
Armageddon!
† These are also found in symbiosis with giant clams, some jellyfish, sea
anemones and some sea snails.

Perhaps the most dynamic example of how this works is found in the fantastical marine lakes of Palau in Micronesia. By best estimates, around 12,000 years ago, at the end of the last Ice Age, rising sea levels filled a basin in the centre of a coral island. Some tiny planktonic larvae were washed through cracks and crevices into the marine lake, which was then almost completely sealed off from the seas beyond. Rotting vegetation formed an anoxic (zero oxygen) poison gas layer in the depths of the lake, making it a lethal environment for small fish that might have been a food source for predators within.

If you swim at the surface, you can witness one of the most surreal sights in nature. In the year 2000, during my tenure with National Geographic, I made a short scramble on a sandy path over the weathered coral rock to reach the centre of Eil Malk island. The lake itself is dark green and flat calm – the rock walls and thick jungle that surround it on all sides stave off even the whisper of a breeze.* I waited for half an hour for a throng of bright orange life jacket-wearing Korean tourists to finish their snorkel and head back to their tour boat. They left behind them an eerie silence; the cameraman and I were to have one of nature's great spectacles all to ourselves.

We didn't wear scuba equipment, as our bubbles could disturb things and potentially stir up the lethal hydrogen sulphide below. We also swam barefoot with no fins to avoid creating any wake that might tear the fragile skirts of the animals within.

Just metres from the shore, I swam from empty green waters into a science fiction movie, with weird planets, stars and alien vessels drifting past my mask. These are the golden jellyfish known as *Mastigias*, swarming here in vast shoals or smacks.

---

* This fluke of geology and flora has probably helped to prevent the poisonous gas layer mixing with the rest of the water, which would have surely killed off everything living within.

Each one is hypnotic in and of itself, with its translucent golden colour; sunlight diffuses through them and their bells pulsate in mesmeric time, driving each individual along. They do not move or pulse in synchrony with those around them, but there is a common aim. Although it was the middle of the day and the sun was directly overhead, the smack was slowly moving from east to west to ensure the jellies spent the maximum time basking in the sun's rays. It is imperceptible when you are in among them, but seen from above, and particularly in timelapse, the whole smack of millions of jellies is constantly migrating around the pond, following the sunlight to fuel their endless summer.

Swimming into the epicentre of the smack, it was truly overwhelming, with so many jellies that every stroke sent hundreds of them swirling. My camera operator was only two metres away, but I couldn't see him, as there were just too many jellies in the way. It felt somehow wrong to be in among them. They are described as stingless jellies, having lost their need to sting, as there is nothing here to kill, but I felt an occasional tingle, which might have been from the remnant of a weak stinging cell or might have been psychosomatic.

Each one of these 10 million jellies is like a mobile coral polyp. Or rather, each coral polyp is like an upside-down and anchored jellyfish. The gut and anus is the same opening, with tentacles that stretch out to capture pieces of plankton from the water column. And deep within their tissue is a hidden secret, the same gift that has been giving to corals for millions of years: the minute algae called zooxanthellae.

The zooxanthellae are dinoflagellate algae that contain chlorophyll, and photosynthesise like plants. However, they've come up with a nice little arrangement where they give energy to their cnidarian hosts. The coral or jellyfish provide safe homes, and the zooxanthellae give them up to 90 per cent of the energy they need for their daily lives.

Organisms within the group exhibit two basic lifestyles. The medusa form is free-swimming, and in some cases (like the deadly box jellyfish) is powerful enough that it can swim against current and winds. Some of these have statocysts like squid, which help them to orientate themselves in three dimensions. Then there is the sessile (meaning they move little or not at all) polyp form, which generally rests on the bottom.

Many of the Cnidaria – corals being the classic example – will have a planktonic stage that is free-swimming, but then will settle down into their colonial phase, in which they might live for centuries. In some situations they reproduce by asexual budding, when a parent polyp reaches maturity and divides off, or they can reproduce sexually, in what is known as broadcast spawning. The stony corals eject millions of gametes (eggs and sperm) up into the water column, which then join together and float about until a suitable substrate is found to set up shop. Once there, the familiar stony coral head starts to grow. They then secrete a hard calcium carbonate skeleton beneath them. Some 'fast'-growing corals, like the branching corals *Acropora*, might put on ten centimetres of skeleton a year. Other species such as brain corals will put on less than a centimetre a year. This means that significant barrier reefs, such as are found in Australia or Belize, could take anywhere from 100,000 to 30 million years to grow.

• • •

When asked what the greatest experience in his unparalleled career was, Sir David Attenborough, without hesitation, said: 'The single most revelatory three minutes was the first time I put on scuba gear and dived on a coral reef.' It is difficult not to agree with the great man.

Much of the wonder Attenborough described comes from the sudden ability to move 'in three dimensions', as he put it; the freedom of soaring over a coral reef like a bird on the wing. Trying to sum up the sea from the surface is like describing the magic of Disney from the theme park gift shop. The key is what lies beneath the waves, and humans have been plunging to the depths since long before the advent of modern diving technology.

For 3,500 years, humans have harvested the treasures of the deep, freediving for precious sea gems like pearls, using nothing more technical than a rock to carry them down. This is apnea, from the Greek for 'without breathing'. In Rome and ancient Greece, freedivers were even used during wars to perform acts of sabotage on foreign fleets.

On exploratory expeditions in the Mexican Yucatan, my team and I found evidence that at least 3,000 years ago the ancient Mayans would breathhold dive down into the cenotes – caves filled with crystal-clear water – to present sacrifices to the gods. The sight of grinning skulls – some of them from children, others from adults who had had their chests cut open and hearts removed still beating – is one that will stay with me for the rest of my life. National Geographic explorer Guillermo de Anda was my compatriot on my expeditions into the sunken cave systems, and explained how the remnants we saw there could only have been swum down and placed in their final resting places by human freedivers.

The Ama in Japan are some of the most legendary of the traditional breathhold divers. Practising ancient arts for extending their depth and bottom time, these 'sea women', who dive for seafood and pearls, have long been eulogised as keepers of freediving lore. Now, though, there are only a few octogenarian deep divers continuing the tradition, which it sadly seems will soon die out in modern-day Japan.

But perhaps the greatest living freediving culture is found among the nomadic Bajau peoples of the Southeast Asian seas, who would once have lived almost all of their lives in small boats out in the South China Sea. When I made my first expeditions into Bajau territory in 1990, most of the 'sea gypsies' were settling into stilted fishing villages in Indonesia, Malaysia and the Philippines, their floors hovering over the shallows inside protected coral lagoons. A very few still lived those old nomadic lifestyles, using the villages as stopping-off and refuelling points.

In Northern Borneo I had the privilege of spending some time with a truly nomadic Bajau family. A 15-strong family comprising three generations, they lived their entire lives crammed into a small boat the size of a minivan. Everything they needed came from the sea, apart from fuel, for which they traded dried fish, sea cucumbers and stingrays. The entire deck and roof of the boat was covered in drying fish of every species imaginable, and inside babies bounced in makeshift cradles over sleeping granddads and piles of live urchins and sea snails. It was the most intense but joyous family scenario I've ever witnessed. Imagine spending your whole life in the kitchen with your parents, grandparents and your kids. Even more weirdly, one of the boat women suffered from seasickness!

The men of the boat garnered all their catch by diving, dropping over the side and down to the bottom for remarkable lengths of time. There was no modern equipment at all. Just wooden goggles, bare feet and one long, deep breath.

In the Mediterranean, people have dived for sponges and clams since prehistory. In the Dodecanese islands in the Aegean, they'd gaze down into the depths through a long pole with a glass bottom. When they found a sponge bed, naked divers would drop 30 metres to the seabed, weighed down by heavy rocks. Their target was the silicate skeletons of sponges, which could be dried out and then sold at a high price for personal hygiene and padding out military helmets. This industry continued until the 1980s, when synthetic sponges took over and a bacterial infection wiped out many of the natural sponges.

The first deep freedive ever recorded was in 1911 by a sponge diver called Haggi Statti in Greece. A boat had lost her anchor and its owners offered a reward for its return. Despite ill health and apparent frailty, Haggi dived down to 77 metres, passing a rope through the eye of the anchor to retrieve it. His reward was £5 and the privilege of being allowed to fish with dynamite. It was many years before Haggi's superhuman feat would be beaten.

The modern sport of freediving, which began in the 1940s, saw great leaps forward in the 1960s with the legendary competition between Jacques Mayol and Enzo Maiorca (dramatised in the Luc Besson classic *The Big Blue*). These two men pushed one another to expand the boundaries of what was then possible, extending records further and further, and dicing with death in the process. Mayol became the first person to break the 100-metre mark in 1976, using a technique called No Limits, in which the diver descends quickly on a weighted sled and returns to the surface using a balloon filled with air.

In 2014, the freediver William Trubridge also passed 100 metres in depth. However, by then the bravado and unnatural techniques of No Limits were long out of fashion. Trubridge achieved his ludicrous feat in the category of constant weight, no fins. That is to say he went all the way down and back up again, swimming breaststroke, without even the benefit of flippers.

Diving using rudimentary technology has a longer history than most realise. Pioneers in aquatic exploration were as daring as the first humans to venture into space. In the 1500s, Leonardo da Vinci drew designs for a deep-sea diving suit that could be used to descend to the bottom of harbours and cut holes in enemy ships. His drawings show a bug-eyed beast with tentacles of bamboo and leather, which looks for all the world like a monster from *Doctor Who*. Like his helicopter, the dive suit wasn't ever made or used, but it is freakishly prescient considering the tech that was to come.

Much modern professional diving takes place on 'closed circuit' or rebreather systems, where exhaled oxygen is passed through a scrubber of lime in order to remove carbon dioxide. These are considered to be cutting-edge technology, and perfect for wildlife encounters as you can stay down longer and there are no exhaled bubbles to scare off the fish. However, rebreathers actually pre-date scuba. The first record of such a system, designed by a Dutchman called Cornelis Drebbel in England, dates back to 1620. Drebbel's wooden submarine made oxygen by smouldering saltpetre in metal pans and contained scrubbers of lime. The idea of going down in a wooden sub, stoking flaming metal pans and having it filled with a substance that turns into caustic acid if it gets wet fills me with horror, and it's fair to say that if I'd been around at the time, I would not have been a diver!

In 1715, Englishman John Lethbridge converted an oak barrel hooped with iron into an underwater diving machine, adding leather bindings to form armholes and a glass window to look through. It was able to contain around 30 gallons of air. He did his first test runs in a well at the bottom of his orchard, before managing to take it down to the bottom of the River Dart to nearly 20 metres for 45 minutes, an achievement which takes the kind of guts I cannot even contemplate. He then invented a sort of diving engine for salvaging wrecks, with air refreshed at the surface through a pair of bellows. This was a sour success, as his cousin stole the patent and 'both honour and profit'.

Then in 1772, a Frenchman, Sieur Fréminet, went down to 20 metres in a leather suit with a copper bell helmet and on his back a cylinder of compressed air. He gave it the catchy name of the *machine hydrostatergatique*.

Before the self-carried units we use now, surface-supplied systems seemed to be the future of diving. The iconic spherical deep-sea diver helmet you'll see represented on logos, tattoos and amulets on modern divers worldwide is the Deane helmet. First patented in 1823, it was originally intended as a 'smoke helmet', created by the so-called 'infernal diver' John Deane with his brother Charles.

The story goes that John was returning from work one day and saw a barn ablaze, with terrified livestock whinnying and braying inside. With barely a thought for his own welfare, John plunged inside with a bucket on his head to protect him, and heroically saved the farmer's prize bull. Inspired, the brothers came up with the Deane helmet, which is spherical, with a number of barred windows. It functioned much like an upturned bucket on the head, using an air compressor to fill it with breathable gas. It failed completely as a fire service tool, as it was so heavy that the fire would have been out by the time the helmet wearer managed to get there.

However, the Deane was soon used on board a boat with a tube that ran down to a diver on the seabed. The brothers used it for their first dive in 1829 on a wreck off the Isle of Wight. It worked pretty well as long as they remained upright – if they leaned forward too much it would fill up with water. Later on, the brothers became legendary for using the helmet when discovering the wreck of the *Mary Rose*.

In the late nineties I used a surface-supplied system like this when diving with Indian fisherman who were gathering giant green-lipped mussels from the seabed. They had a diesel engine on board their rickety rowing boat, driving a pump, which drove air down a garden hose. Underwater they used a leaky wooden mask, and put the other end of the hosepipe into their mouths. They invited me to have a go, and it was one of the most dangerous and dumb things I've ever done. It was hugely claustrophobic, you could taste dank petrol with every breath and there were no protocols for surfacing slowly or doing safety stops. On the bottom I stepped on a sea urchin and spent the rest of the day picking spines from my feet. I also got dysentery from eating the mussels. The Indian divers who did it day in day out had all suffered bends and life expectancy was low.

The first successful scuba dive was undertaken in 1831 by engineer Charles Condert, who created a system with a rubber mask and a pump that he hand operated underwater. He dove it to the bottom of Brooklyn's East River and miraculously returned alive. Knowing that foundling New York had a somewhat rudimentary sewage system, I can't imagine his exploits rewarded him with much of a view – and probably gave him a storming case of dysentery. Sadly in August of the following year the device failed, and he never returned to the surface, sinking to be buried at the bottom in unimaginable sludge.

Proper revolutions in underwater breathing technology came with naval lieutenant Jacques Cousteau's developments of the aqualung in the early 1940s. In tandem with engineer Émile Gagnan, he perfected earlier designs, most importantly adding a demand regulator to their systems. This meant that compressed air could last far longer, as it wasn't just constantly passing away into the water, enabling them to explore further, and finally to start making the first underwater documentaries, paving the way for the thousands of aquatic films that followed.

• • •

Jacques Cousteau drove our understanding of and desire to explore our seas. His seminal film and book *Le Monde Du Silence* remain the most important marine masterpieces of all time, but they have one fundamental very wrong. Our seas are not silent. Far from it.*

Sound travels further and faster underwater. Science has known for decades that whales and dolphins operate in a whole different world of sound to us, with hearing being their most important sense. It is, however, a revelation how much everything else in the ocean is dominated by sound. Coral reefs have always been portrayed as a visual feast. Even natural history classic *The Blue Planet* showed reefs devoid of natural sounds. This might turn out to be a TV faux pas up there with pushing lemmings off a cliff.†

When the BBC made their follow-up, *Blue Planet II*, they employed Professor Steve Simpson of the University of Bristol to set that right and provide them with the appropriate reef soundscape. Steve calls himself a marine bioacoustic nerd. As fit as a butcher's dog, with limitless youthful enthusiasm and the great gift of being able to make even the most mundane topic feel like a detective story, Steve is the lecturer every young scientist should have.

The central focus of his work is reef bioacoustics, figuring out the sounds of reefs and their residents, as well as how sounds could be used to give a helping hand to our ailing marine treasures. For a few privileged weeks in 2022 and 2023 I got to be his research assistant.

On the wooden dock of a Maldivian dive resort, he mounted a 360-degree camera on a tripod, able to record vision in every direction. We then took the camera down to the middle of a

---

* In fairness, Cousteau dived on extravagant rebreathers fashioned from the scrap heap of local fire-fighting mechanics, and usually in chain mail as he was so scared of sharks. So he probably only heard clanging and roaring bubbles!
† The 1958 Oscar-winning Disney documentary *White Wilderness* showed migrating lemmings leaping off cliffs to their doom. It turned out that they'd been chucked off by filmmakers eager to realise a (falsely) reported behaviour.

flourishing coral reef and surrounded it with a circle of sound recorders. This array was left in place for a few hours to do its job. Steve then took four waterproofed microphones, a dive float, three broom handles and some polystyrene bowls, and fashioned a multi-directional hydrophone array he called 'the hammer-head' for its vaguely H-shaped profile. We took this preposterous and fragile monstrosity an hour by boat to 'Hassan Haa', a reef named for the Maldivian explorer who discovered it.

These Indian Ocean reefs have suffered hugely from bleaching events in the last few decades, but the one we were now swimming above was the most perfect I had seen since the Millennium. The small upright bushes of *Pocillopora* in cream colours with pinky purple tips were bunched alongside the comparatively fast-growing branching *Acropora* like the antlers of a hundred deer clustered together. And bulging up between them all, the pleated domes of the brain corals, which is what you would call them even if you didn't know that was their common name.* These massifs are some of the slowest growing and can get to be 2,000 years old, and as their polyps are connected and not separated like those of the branching corals, they can share nutrients, hormones and oxygen.

Shoals of black, white and yellow stripy sweetlips glided by in unison, and herbivorous unicornfish with their extravagant forehead horns nibbled algae off the corals. I also saw the biggest shoals of clownfish I have ever seen – in some cases hundreds of colour-crazy fish inhabited clumps of anemones covering automobile-sized coral colonies.

Steve and I swam at the surface above the coral metropolis, dangling the parabolic mics below us on ropes, as if we were

---

* The Maldives announced in 2020 that they would be building a floating city for 20,000 people outside of the capital of Male, designed around and in the shape of a massive brain coral.

suspending a rather shoddy chandelier under a zeppelin. It was by far the most ungainly and frankly Heath Robinson bit of science I've ever been involved in, and stood no chance of garnering any results. Or so I thought.

That evening we sat down round Steve's laptop with a cold beer each and ran back through some of the results. Astonishingly, the hydrophone array had recorded a symphony of sound. It sounded a bit like a cross between popping corn and the clacking of pool balls after a particularly enthusiastic break.

'Those are the sounds I associate with reefs,' I said. 'Most of those are from pistol shrimp and parrotfish, right?'

'Correct,' Steve said. 'The parrotfish crunch the coral heads, and digest the living material. What we're hearing is their beaks and teeth crunching up rock. Then the calcium carbonate comes out as pure white sands.'

Parrotfish teeth are made of a fluorapatite, which is the second hardest biomineral in the world, harder than most metals. A big parrotfish could excrete a tonne of sand out in a year.

'Which explain these glorious white sand beaches?'

'Right. But the snapping shrimp is even more interesting and important.'

The alpheid shrimp – despite being able to comfortably sit in a human hand – competes with animals like the sperm whale for the title of loudest animal on earth. Most dig burrows* and are found not only on reefs, but in seagrass meadows and sandy seabeds. Their dramatic sound is not a call as such, but a function of their oversized claw, which might take up half the size of their body. Rather than having pincers on it, it has an action more like the firing hammer on an old-fashioned pistol. When the claw 'fires' shut, it emits a wave of bubbles that then collapse

---

* Which are shared in a symbiotic relationship with goby fish, who have better vision and function as sentries.

in what is known as cavitation. This precise moment has been captured on high-speed film, and you can see a white hot flash at the epicentre. For a millisecond, that cavitation bubble is as hot as the surface of the sun, and the sound that it emits can smash crab shells, disorientate fish and carry a great distance.

'So far, so familiar,' Steve said. 'But there is actually an awful lot more going on than we ever realised.'

At this point Steve switched to the 360 camera. He had synced up the audio devices to match what we could see. In front of the camera was a large coral head, which had a substantial anemone in the middle, filled with flitting clownfish.

'OK, so let's listen for something different now,' Steve said.

No more than a minute or so into the recording, there was a sort of burping sound, then bubbling, like when you listen to someone's blood pressure releasing through a stethoscope.

'There!' I said. From the corner of frame, a small wrasse swam into view. One of the clownfish darted out to face him, and at that exact moment we heard the sound.

'Yes!' said Steve. 'That's it. So what we're hearing is an alarm call or threat display from the clownfish.'

'How does it make the sounds?' I asked.

'Lots of different ways,' Steve replied. 'In this case it's probably a vibration of the swim bladder.'

I nodded. The swim bladder is a gas-filled box which many fish species use as a way of regulating their buoyancy. Diving in a reserve off Cuba, I was hand-feeding a goliath grouper the size of a fine farm pig when a silky shark took interest in the bait and got too close for the grouper's liking. The grouper flexed its whole body and twanged its swim bladder with a sound like a pluck of the deepest timbre string on a double bass. I felt the pressure wave through my whole body. The shark sped off like it had been zapped with a cattle prod.

'And don't grunts and sweetlips grind their teeth?' I asked, instinctively followed by a shudder. My little boy grinds his teeth in his sleep – the sound of fingernails scraping down a blackboard is like the music of an angel's harp by comparison.

'Or it could be a click of the jaw,' Steve continued. 'Or they can be popping synchronised bubbles out of either end.'

'Attack farting?'

'Or burping, yes.'

Fish flatulence almost created a diplomatic incident in Sweden in the 1980s. The Swedish navy's underwater listening systems were detecting huge volumes of unexplained aquatic noise. Surely the product of espionage, it was thought – most likely Russian submarines penetrating into their sovereign waters. Imagine their relief/embarrassment when it turned out the sounds were actually herring farting in unison (probably as a means of communication).

'The amazing thing is, though,' Steve said, 'this clownfish is unique to these reefs, and no acoustic work has ever been done here before ...'

I filled the empty space: 'So we are listening to their vocalisations for the first time?'

'That's right. No one has ever heard – or at least recorded – the calls of the Maldivian clownfish before. This is a world first.'

As we watched the videos we'd recorded, we saw half a dozen fish interactions in among the coral heads, all of which were accompanied by a vocal cue of some kind. At one point a regal angelfish swam into shot. It is a glorious fish, with a dark blue mask and blue and yellow longitudinal lines running down its flattened advertisement board of a body. As it swam into an occupied hidey hole, a crystal-clear 'WUP wupwupwup' sound emanated to the camera.

'There it is!' Steve pointed jubilantly. 'The first ever recording of the regal angelfish.'

In the coming months, Steve and his acolytes returned to the same reefs and recorded different sounds at different times of day. At one point he saw and swam after another regal angelfish. As he approached it, Steve heard that same distinctive call, and through his regulator mimicked the sound. The angelfish promptly called back to him. Was this the first ever conversation between a person and a fish?

Steve's work is still relatively new, and it's painstaking.

'This work can only be done in natural environments,' he told me. 'In captivity in aquariums they don't sing or call. Probably because they don't have natural territories and are not undergoing natural breeding behaviour.'

Steve and colleagues around the world have been busy recording sounds for over a decade, and it is showing that each fish species has its own unique call. 'We've seen that fish which are nesting are much more likely to defend their patch with sound,' he said. 'And not only that, but just like birds, they may have a song to attract mates, a subsong, contact calls, startle and alarm calls …'

It's an extraordinary thought. The songs of Britain's 630 bird species have been known to any enthusiast for many hundreds, and potentially even thousands, of years. But there are more than 33,000 species of fish, and most have never had their vocalisations recorded. Steve and his fellows have logged a few calls from just 1,500 species. The fish have always been talking. We just haven't been paying attention.

'We're in the era of citizen science,' said Steve. 'I just love the idea that anyone could just get a GoPro on a stick and go and describe for the very first time a fish's song. These are real discoveries that anyone can make and record.'

A whole new field of zoology is being born. Every single fish we saw on this reef, plus those under Arctic ice and in the deepest oceans, even the pike and perch in the Thames at the bottom of my garden, is communicating in ways even the most obsessed marine biologists have never even contemplated. And a kid with a mask and snorkel could potentially drop into a body of water and come out with a world-first discovery, purely by using their ears.

This is just the tip of the iceberg. The most exciting thing about Steve's work is that the overall ambience – the orchestra, if you will – of a coral reef is markedly different on a healthy one to a recovering or denuded reef.

A year later, in 2023, we returned to the Maldives to continue the work, but this time with the addition of another world-renowned scientist, Professor Peter Harrison of James Cook University in Australia. Peter is another legend in his field, and at the heart of ground-breaking research when it comes to coral reproduction. It's fair to say that if I hadn't been intimidated before, I was now!

Along with Peter and his team, we went out at night three days after the full moon in March and were witness to the spectacle of synchronous coral spawning. We sank slowly down into shallow waters, the coral heads stretching out around us like vases, baskets, statuettes and well-heeled old leather footballs. Our lights were dulled with red gels to avoid putting the lovers off their courting. Predatory black jacks rocketed in and out of the crannies and nooks, and huge-eyed red soldierfish lingered in alleyways. The brightly coloured aquarium fish of the daytime were all gone, vanished into their midnight bowers. It was two nights after the official full moon, but the ripe yellow belly of the lunar orb still shone silvery light over the waves above us.

Around me half a dozen lithe marine biologists in overlong freedive fins plunged among the formations, placing gamete traps over individual coral heads – effectively coral condoms that would catch the sexual cells as they emerged. Corals can reproduce in two ways. A colony can be male or female, in which case one will release eggs, and the other the smoky slew of sperm that will fertilise it. Otherwise the colony can be both male and female, and will instead release packets of both sets of sex cells together. On cue, members of the same species will ease bundles of eggs and sperm towards the surface of the coral heads. They then release these bundles en masse. A small blob of fat inside them makes them buoyant, sending the shiny pink baubles to the surface in squalls – it's rather like being inside a snow globe. Peter's work was to trap and protect these sex cells in order to increase their odds of survival.

Four days later, we pipetted out tiny planktonic coral larvae and looked at them under a microscope. They were elongate lumpy cigars but moved with determined force. It was a strange thing to watch, as they were clearly swimming, but you couldn't see how until you increased the magnification. Then you could see they were using whiplike flagella called cillia on their surfaces to swim, driving themselves across the surface of the water. Steve described them as looking like fuzzy tennis balls after a game in the rain.

'Until very recently,' Steve explained, 'We thought these planktonic larvae just floated around, and then eventually sank to the bottom. If they were lucky they fell in a good place, but if not, they just died or were eaten by something. But when you look more closely, it's nothing like as random as this.'

It turns out that the larvae will orientate themselves towards existing coral reefs, giving themselves a better chance of landing somewhere perfect to begin their lives. The cues that drew them in, however, have been a mystery. Until now.

Steve placed one of his audio recorders into a tank within which was a perspex tube containing swimming coral larvae. To either side of the tank were speakers.

'So over time,' Steve said, 'we've managed to isolate the soundscape from different times of day. The midnight sounds are totally different to the middle of the day – there are still loads of snapping shrimp but no parrotfish scraping around! This, though, is really cool.' At this, Steve played the richest and most complex reef sound we'd yet heard. Crackling static from tuning an ancient radio, mousetraps snapping shut, burps, chirps and cricket chirrups.

'This is the dawn chorus!' Steve was as excited as a schoolboy. 'Those are the songs of things like damselfish, as they awake from their night tucked up in bed and are now in full voice to reassert their territory.'

'It couldn't be more different sound-wise,' I commented, 'but from a natural history perspective, that's exactly what a birdsong dawn chorus is doing back in a British woodland.'

As soon as he started to play the sounds of a healthy coral reef on one side, the larvae visibly oriented themselves towards the origin of the noises and swam towards them. Their subsequent descent to the bottom was actually a good deal faster than if they were just sinking. They were swimming, propelling themselves down to their new home. Even more extraordinary, when Steve played sounds recorded on a healthy coral reef and those from a denuded one, the larvae chose the healthy reef. The louder sounds of the snapping shrimp hidden away in the nooks and crannies perhaps let them know that there is prime real estate here. The grazing crunches of the parrotfish spell out to the corals that here is neighbourhood that they can make their own. Somehow these microscopic entities that you can only just make out with the naked eye, which lack nervous systems or evident senses, can detect and then find their way towards the best locales, ignoring the dodgy estate agents and letting the buzz of the neighbourhood do the talking.

Steve and his colleagues' research has shown that juvenile fish select reefs to live on based on their sound landscapes. They have even proved that by playing through underwater speakers the sounds of a healthy reef on a unhealthy one, they can encourage fish to make their homes there. This has particularly positive implications, as once herbivorous fish return to a floundering reef, they can start to clean off the choking algae which other-wise will prevent recovery.

'We cannot only get the punters to come to the party,' Steve said, 'but by rocking out their favourite plankton playlist, we can convince them to rebuild the nightclub!'

• • •

The field of coral research is one of the great growth areas of modern marine biology. Some of that is surely down to young biologists wanting to spend their time in the field in paradise-like tropical seas. Perhaps more importantly, though, the coral reef is one of the most threatened environments on earth.

A big part of this is mechanical destruction, in the form of absurdly short-sighted fishing practices. In 1996, working as an author for Rough Guides, I travelled all the way along the Indonesian archipelago, gathering information to fill a guidebook for adventurous backpackers. One of the highlights of the journey was to the far-flung island of Alor, to the east of the archipelago.

In the straits between Alor and its neighbour Solor is a pinnacle – a submerged peak – rent by impossible tidal currents, which come maybe 15 metres close to the surface. Back then, Kal's Dream was a near-mythical site, spoken of by divers in hushed tones as one of the world's great aquatic secrets. I only dived it twice, but they remain to this day the finest dives I've ever done, with swirling shoals of jacks, silvery Spanish mackerel and giant bluefin tuna the size of cattle circling between the sharks, mantas and barracuda.

Eight years later, while working for the BBC's Natural History Unit, a colleague got in touch with me. They were putting together a new mega-series for David Attenborough called *Blue Planet*.

'You've been diving just about everywhere,' she asked. 'What's the best dive site in the world?'

I didn't even pause. 'Kal's Dream,' I replied. Several months later, I received a phone call from the same clearly annoyed researcher. 'We looked into your "best dive in the world",' she said, with the air of a sarcastic copper who'd been told one too many porkies. 'It's hardly worthy of a series like this. It was bombed and cyanide-fished years ago – it's nothing but rubble.'

The tragedy of the Dream has been replicated throughout the tropical coral reef world, but I've seen it at its worst in Indonesia and the Philippines. Homemade bombs, sticks of dynamite, even lethal recovered Second World War ordnance are dropped over the sides of dugout boats into the water, the explosions stunning or killing the fish. Quite apart from the destruction of the environment, the danger to impoverished fishermen is unthinkable. One of our guides in the Philippines was a former dynamite fisherman and only had one arm. The other had been blown off by the untimely detonation of a homemade grenade.

Even more cataclysmic and just as dangerous is cyanide, which is diluted in plastic bottles with a hose at the end. Freedivers or hookah divers (breathing though a hose to a simple compressor at the surface) drop down and pump the milky poison into the cracks in the reefs, in order to kill or stun the biggest fish that live there. Those that are killed are taken and eaten by local people, but some of the large ones survive and are taken to be sold through the insatiable live fish trade, which transports live reef fish all over East Asia.

It probably doesn't need to be pointed out how dangerous it is for people to be eating fish that has been killed with a notorious poison, and untold thousands of families – especially young children – have suffered or been killed by cyanide poisoning. In addition, the half-life (or time it takes half of a substance to degrade) of hydrogen cyanide is between one and three years. The toxin that lingers in the reef kills everything, and even poisons the substrate so the corals can't return. Thousand-year-old coral heads can be completely destroyed in a single incident. I can remember the sick feeling in my guts when I learned about the destruction of Kal's Dream even now. One of the world's natural wonders was gone – nobody mourned and nothing could be done. It was just gone.

Twenty years later, and the Dream is back to being a great dive, but no one will ever experience it the way it used to be. I've seen the same situation repeated again and again in so many of the world's finest dive destinations. Coral reefs that can live thousands of years, and withstand the pounding of hurricane seas and tsunamis, can be killed in seconds by human hands.

The bigger and more insidious problem is down to ocean acidification, especially as a side-effect of climate change. In 2013 we reached more than 400 parts per million of carbon dioxide in the atmosphere for the first time,* and every year brings new and unwelcome records for temperature increase, ice melt and drought. We are now in a situation where climate change is becoming real for all of us.

With even slight increases in sea temperature, corals will expel their symbiotic zooxanthellae, which results in coral 'bleaching'. The white skeleton of the coral remains, but the life has gone. Our oceans have enormous ability to absorb heat, but Lijing Cheng of the Chinese Academy of Sciences has worked out that over the last 25 years our oceans have absorbed the equivalent

---

* For comparison, pre-industrial CO_2 levels were approximately 280 parts per million. Today, in 2023, we're at more than 420 parts per million.

excess heat energy of four Hiroshima bombs every single second. With sea temperatures already up by around 1°C on pre-industrial levels, there have been mass-bleaching events all over the world. Perhaps the best known is the 2016–17 bleaching on the Great Barrier Reef, which killed approximately half of the coral there. Many big scientific groups such as NASA, NOAA and UNESCO suggest that unless we act to radically reduce ocean temperatures there will be no reef-building corals by the turn of the next century.

In the 1980s, when I was a kid, there was a sudden drive to protect the rainforests. One of the main arguments as to why the destruction of the Amazon, for example, was a bad thing was the idea of 'bioprospecting' – that the plants there held the answers to many of our modern needs, including potential cures for diseases such as cancer. It would therefore be disastrous to lose the forests before we'd discovered their secrets. This is also true of our ocean ecosystems.

In the year 2000, I travelled to the Australian Institute of Marine Science, or AIMS, and met Professor Chris Battershill, a Kiwi microbiologist searching for pharmacological compounds from marine sources. His chief subjects were bryozoans, corals and sponges, all of which seemed ripe for research.

These organisms have been around for hundreds of millions of years, and being mostly stationary have had all that time to come up with alternative defence mechanisms. They can't bite back or swim away from predators, but they are far from harmless. Indeed, some species of sponges create a cocktail of 24 chemical compounds if something tries to eat them. Nearly all sponges on tropical coral reefs are poisonous to eat, but the clever bit is that this toxicity is not permanent. It would cost them too much energy to be continually synthesising toxins in their tissues. Instead, the sponges generate these toxins in response to threats or attacks. Simply speaking, a turtle, for example, nibbles on a sponge and within seconds the sponge creates poison in the tissues that are being attacked, giving the poor turtle a mouthful of foul-tasting or even lethal fodder. It would be like me feeling a mosquito biting my ankle and instantly turning the blood in my foot into a poison that would kill it.

These toxins are extremely complex and have been suggested as possible cures for a variety of human ailments. Some of them have anti-tumour properties known as bryostatins. Most toxins that have cancer-killing abilities are so poisonous that they kill everything they come into contact with, but bryostatins appear very low in toxicity, isolating cancer cells and leaving others healthy.

More than 20 years later and animal venoms and poisons remain one of the most exciting areas for pharmacology. My own academic research, started in the early 2000s and finished in 2018, tested the venoms of newts and salamanders on cancer cells* – they did kill the cells, and had a decent anti-bacterial and anti-microbial function; however, they were pretty indiscriminate and killed everything else too.

Other sponge and bryozoan compounds and cells have the capacity to eternally regenerate and metamorphose, and are being touted as possibly providing a source of stem cells that could be introduced to humans to end or stall the ageing process. The theory goes that if these stem cells could unite with human cells that have a limited shelf life, we would never degenerate and could live ... well, forever.

This might sound like science fiction, but there is a cnidarian which can theoretically do just that. The so-called immortal jellyfish goes through the usual cnidarian life cycle, with larvae

---

* I narrowly missed out on discovering the first example of an amphibian venom by a few months. Instead I had to make do with the first venom from a newt or salamander (collectively known as the Urodela).

forming into mature adults. But then they reverse this process, going all the way back to being a juvenile before repeating the process ad infinitum. In the wild these jellyfish would inevitably succumb to predation at some point, but in the lab scientists like Chris have shown them to be functionally immortal.

Chris's task is to find a way of rearing/farming these organisms, and then trigger the specific defence mechanism that will initiate the creation of the compound he requires. This can be as crude as taking a small chunk out of the sponge's tissue to simulate an attack by a predator or something more subtle, such as introducing chemicals that mimic the activity of parasitic barnacles. The end game may be the end of cancer, of disease, of death, and presumably a Nobel prize at the end of it. However, the process is incredibly labour intensive and carries with it great expense.

As Chris took me around the labs and exhibits of AIMS back in 2000, he showed me a cold storage unit the size of an average family chest freezer, where samples were kept at $-86°C$. Within were samples of sponges that would once have looked to me just like a normal bath exfoliator, but now appeared as the collagen skeletons of noble organisms, one of the first multicellular beings to evolve on the planet. The extensively purified compounds in this small unit alone were worth tens of millions of dollars.

A few months after I got to be Professor Steve Simpson's research assistant in the Indian Ocean, I was invited to open an ecological centre in Laamu Atoll, with the carrot of bringing my young family out to experience a coral reef for themselves. The flight with three-year-old twins and a four-year-old was pretty torturous, and seeing them bamboozled by jetlag and lack of sleep, I thought maybe I'd made the wrong call to bring them. Once settled on the atoll, though, the giant swooping fruit bats, scuttling agamid lizards and scampering ghost crabs with their eyes on comedy stilts delighted the three of them, and soon we relaxed enough to let them run feral around the island.

We had been careful not to put too much pressure on Logan to snorkel over the nearby reef (young kids having a tendency to do the exact opposite of what you want them to), but as soon as he saw the aquamarine waters, he was begging us to give him a mask. I have never seen such pure unbridled joy bubbling out of a human. By the end of the week, he had spotted Nemo and venomous conefish, had hawksbill turtles surfacing right by his face, and seen half a dozen species of sharks and rays. Even better, the twins, on sensing his joy, decided they had to join him, and, still wearing their swimming pool armbands, they spluttered above clownfish colonies and moray eels, staying in the water until they had to be physically dragged out. My little girl Bo has an utter fixation with unicorns, and when she saw her first unicornfish up close, her screams of joy could have been heard a mile away. It was without doubt the highlight of my – as yet short – parenting life. (Rather less so when Bo reminded us of why three-year-olds probably shouldn't snorkel all day by projectile vomiting seawater all over the dinner table.)

Between 70 and 90 per cent of the world's coral reefs will disappear within the next two decades. Most predictions suggest my kids' kids will never see one. There are so many reasons that this cannot be allowed to happen – for the health of our seas, the protection of our land and the state of our fish stocks. But to me right now, the most urgent is that no other place has the power to instantly create a nature lover or a conservationist than a tropical coral reef.

CHAPTER ELEVEN

Gentle Giants

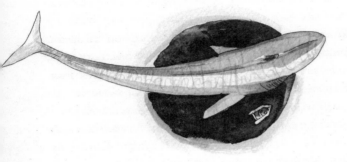

I must down to the seas again, to the vagrant gypsy life,
To the gull's way and the whale's way where the
wind's like a whetted knife;
And all I ask is a merry yarn from a laughing fellow-rover
And quiet sleep and a sweet dream when the long trick's over.

John Masefield, *Sea-Fever: Selected Poems*

The Indian Ocean swell slaps against the side of our wooden schooner, its spume not sufficient to drown out the noisy vomiting of our researcher, who has been kneeling at the rails ever since we set out that morning. She has gone from yellow to green and now grey. Every human instinct is that we should return to shore before she vomits up a spleen, but this is our final day at sea before the storm hits in all its intensity, and our target will be gone for ever.

And what a target. The biggest animal ever known to have lived on our planet. Imagine a 767 without its wings, and you have both the dimensions and weight of a female blue whale. It is the leviathan of legend, and nothing prepares you for the sight of one. Is it banal to comment that a blue whale is really ... blue? Intensely, obviously blue. On other expeditions where I've sought blue whales in the Arctic, Pacific and Sea of Cortez, countless big whales have popped up nearby, and I've thought ... 'Is that one?' Now, however, I know. If you're asking yourself that question, then the answer is no. When you see a blue, it is unmistakable!

And then a spout. A giant, billowing two-storey-tall spout that could be seen even if the animal was beyond the horizon. I slip on my freediving fins, swing my legs over the side of the schooner and wait. Diving with blue whales is not like swimming alongside humpbacks or sperm whales. While some of the more cerebral cetaceans might be intrigued by the strange intruders in their world and stay to play, blues never are. They just motor past, never seeming to stop moving, and avoiding anything they come across in the water – even seals and dolphins. Truly the biggest leviathan on the planet seems to me to be the most nervous of all whales.*

In our three days freediving in Sri Lankan seas, we've not seen anything more than a fleeting glimpse as the blues power past us and off into the distance. It seems as if the local motor boat drivers are too keen, revving after the animals too hard, keeping them on the move. After the end of the second day I organise a team powwow and request a more softly, softly approach. Can we try to just sit tight, put the engines in neutral and slip into the water quietly if one comes nearby? The boat captains look at me as if I'm doolally, but agree. The next morning I get one and only one chance to try out my theory.

---

* Some might say this is down to the persecution that allowed us to remove more than 95 per cent of blues from our oceans. I personally think they are relatively low in intelligence, with a simple feeding strategy of swimming with an open mouth into vast aggregations of food. We're just not of interest to them! I've seen identical reactions from sei and fin whales, related giants that live similar lives.

As we sit, engines off, trajectories align, and the vast plume of a spout bursts nearby. Barely daring to draw a ripple, I ease into the water. Instantly, there it is. A dark, vast submarine coming out of the clear, clear blue. It is so much thinner than you'd imagine, and with a blunted end to its head – nothing at all like the immense inflated models you see at natural history museums.* As it heads towards us, I take a chance. Gulping desperately, I jacknife at the waist and dive straight down, powering blind down to 20 metres. As I turn in the direction I think it will be coming from, a silent freight train, dragging the very water around it along, pulls me into its vortex. And then the vast tail flukes sweep past. They are as wide as a bus. The swirl tosses me like an autumn leaf in the breeze, a mere piece of plankton alongside a giant.

. . .

Today there are 90 species of whales, dolphins and porpoise in 14 different families, all in the scientific order of Cetacea, known commonly as cetaceans, which comes from the Latin *cetos* and the Greek *ketos*, both of which simply mean 'whale'. All the animals share in common a spine that articulates up and down, flattened foreflippers and absent hind limbs (though there may be vestigial remnants beneath the surface). They fall into two main groups: the toothed whales (odontocetes) and the baleen whales (mysticetes), which refers both to their anatomy and how they feed.

The baleen whales are (with the addition of the toothed sperm whale) the biggies, often huge-bodied, and with layers of whalebone or baleen hanging from their upper jaws. This substance is kind of like a hairy sieve, used to filter small fish and crustaceans

---

* This representation is not necessarily false, but shows them full of water and food as they gulp feed, with their bodies hugely distended. When in motion, they are laterally compressed, extremely hydrodynamic, and arguably the fastest of all whale species.

like krill out of the water. It was also used by Victorians to give volume to their corsets.

Feeding in the baleen whales is not a passive process – these are targeted and often coordinated predators. There have been very few times that I've seen a baleen whale lunge-feeding underwater, and every time it has been awe-inspiring. A humpback will power towards its food in its most streamlined form, then at the very last second it drops its lower jaw, allowing its pleated throat to expand, dragging in tonnes of seawater and everything in it. Drag stops the animal dead, the mouth closes, the gigantic tongue (which can weigh as much as an elephant) presses the water out and the baleen sieves out the food. Some blue whales will spin 360 degrees in a pirouette before feeding – no one knows why.

Baleen whales have two blowholes, and on some species (such as the so-called right whales) this can be clearly seen in their two distinct spouts. Female baleen whales are the biggest animals we know to have ever existed on our planet. (I use the qualifier carefully, as there is always the possibility that there were extinct animals even bigger; we just haven't found them yet!) They live for a long time and undertake epic migrations that cross the planet. The humpback is the longest migrant of any mammal, and the bowhead whale (which only lives in Arctic seas) is the longest living. In 2007, a bowhead was caught in Alaska with an unexploded harpoon head buried in its blubber. The type of explosive was only used in the 1880s. As the whale would probably have been mature when it was first hunted, this suggested it was a prodigious age. Other specimens have been aged at well over 200, with their genome sequence leading researchers to estimate that they could reach 268 years of age.

The toothed whales are far more diverse in form and function, from the tiny vaquita to the weirdness of the strap-toothed and ginkgo-toothed beaked whales and the ludicrous narwhal. They tend to feed at deeper depths than baleen whales, and use the large fatty protrusion at the front of the skull (known as a melon) for their hyper-developed echolocation. They have one single crescent-shaped blowhole, and also strangely lack olfactory lobes, suggesting that smell is much more important for baleen than for toothed whales.

The evolution of whales and dolphins is bewildering and fascinating, undoubtedly among the most unlikely of stories in the natural world. They are among the few mammal species that are tied to the sea for their entire life cycles, and yet they are descended from a landlocked ancestor. Sharing a common ancestor with the hippo, this and other early whales were descended from terrestrial mammals, who at some stage ventured into the seas and never came back out again. Some still retain a sprouting of fur around the lips at birth, and they suckle their young with milk and have to return to the surface to breathe air. There are familiar mammalian signatures under the skin too: the front flippers are essentially big hands, with the same bones you or I have in our fingers. Some species retain relic remnants of a pelvic girdle. The articulation of the spine is also a tell, flexing up and down like that of a galloping horse, rather than from side to side like a fish. In order for this movement to fulfil its function in water, cetaceans have tail flukes that are horizontally orientated as opposed to the vertical tails of fishes.

The fossils of early proto-whales are well represented and tell a tale of an animal that lived an amphibious existence. One of the earliest is *Pakicetus*, a fossil of which was found in 1981 in what is now mountainous Pakistan, but 50 million years ago was once a group of islands, where this low-slung, probably predatory beast could have moved in and out of the water in search of prey. Rather than being paddle-like, its limbs were legs that could have supported its weight out of water. In size it was no bigger than a border collie. *Basilosaurus* was more recent, and far bigger, approaching almost 18 metres in length. It looked like a swimming nightmare, with teeth spilling from its reptilian maw. In these early species the nostrils were still at the end of a vaguely horse-like snout.

Maiacetus (the mother whale) moved and looked quite like a sea lion and is best known for one remarkable fossil, which shows a female in the process of giving birth. The calf is orientated head first, which may show that *Maiacetus* gave birth on dry land (modern cetaceans give birth tail first, so the calf has access to oxygen from the mother for longer before birth). *Maiacetus* was the first whale ancestor to have an unusually flexible spine, losing the fast connection to the pelvis, which is hypothesised to point towards them using their tails for propulsion.

Ambulocetus or the walking whale wasn't discovered until 1994, roughly around the same time that Blur and Oasis were battling it out for chart dominance. It was perhaps more crocodile-like in appearance, with eyes and nostrils situated on top of its skull, so the body could be immersed while the senses were exposed above the water. It might have spent even more time in the water than *Maiacetus*, though it still had the ability to support its weight on land. The tail vertebrae share similarities with modern whales, with a structure that suggests they supported fluke-like tails.

None of these animals look even remotely like a modern cetacean, so how then do we know that they are of the same lineage? Well, a lot of that is down to the inner ear bones, which fossilise readily, and show marked similarities to modern whales. Notably, on the outer ear bone sits a thickened bony feature called the involucrum. Modern whales use it as a connection to the inner ear bones and the jawbone, both of which are used to pick up sound in the multi-directional medium of water.

It was probably around 40 million years ago that the ancestors of whales lost their hind limbs and ability to support themselves on dry land, and moved their entire lives into the sea. Somewhere between then and now, the nostrils continued their migration towards the top of the head.

Also during the last 40 million years, some whales evolved their baleen and the ability to suck, lunge or filter feed. One idea is that 34 million years ago, in the Oligocene, the world's seas started cooling down and the precursor to the modern Antarctic current appeared, with Antarctica moving away from South America and Australia. This started to pull up nutrients from the bottom, increasing the productivity of the waters in the form of plankton, and expanding the food chain as a result of this prodigality. Suddenly it was advantageous to bulk feed rather than pick your food with care. Any animal that could filter some of this enormous bounty was onto a winner.

Early mysticetes might have used their teeth like a sieve (a bit like modern crab-eater seals do today) but then baleen evolved as an outgrowth of whales' gums (today's Dall's porpoise has keratinised gums that could be a stepping stone towards this). Meanwhile, the toothed whales went down a totally different route, developing that dome-shaped melon at the front of their skull, phonic or monkey lips to generate sound, and the ability therefore to echolocate.

. . .

In some ways studying the whales of the past is easier than studying their modern descendants, the big problem being that they appear to be avoiding us. For the majority of human history, we have only seen them when they surface, which is a tiny fraction of their lives. Most whales appear to treat their time at the surface breathing as an annoyance, wanting to get back down as soon as possible. Our early whale records were a result of the observations of the behaviour of terrified animals escaping pursuit from whalers' harpoons, and from the necropsies of those same whalers as they assessed their catches. This is like trying to understand people by just looking in warzone mortuaries.

Land-based biologists take for granted that they can, for the most part, watch at their leisure their subject animals doing stuff: eating, flirting, mating, dying. Many elements of what we call 'life history' were first observed and recorded by early scientists thousands of years ago. It is then breathtaking to note that no one has ever seen most species of cetacean doing the most basic things like mating or feeding. It was not until the advent of 'critter cams', 'snot bots' and satellite transmitters that we've even been able to begin to poke into their actual lives.

Modern pressure tags recently showed that Cuvier's beaked whales might exceed the sperm whale in its dives, heading down as far as 2,992 metres in depth, on dives lasting up to 222 minutes. However, no other beaked whale has been tracked in this way. Deraniyagala's beaked whale has never been seen alive, let alone studied, and it is only known because of seven dead specimens that have washed up on beaches.

The next decade of study of cetaceans represents potentially the most exciting field of whole-organism biology in our planet's history. Scientists studying cetacean communication have recently discovered that some species use names for themselves and others. Those same scientists are promising that we will soon be able to decipher what whales are 'saying', and potentially even to talk back to them. I find this possibility entrancing, especially as some of my most potent interspecies interactions have been with cetaceans – even without a common spoken language.

· · ·

The humpback is the marine creature I've worked with the most, from the high Arctic to Antarctica and a hundred places in between. This is down to the fact that despite humpbacks not being the most common whale species, they are the most conspicuous. They don't disappear from view for hours on end like sperm whales do, and they don't power off into the distance like a blue, fin or sei will. Instead they tend to breathe easily and obviously at the surface and then dive for no more than 20 minutes. Every whale watcher, photographer and marine biologist knows and loves its curving notched back as it crests the water to breathe, the giant Y-shaped tail flukes lifting a salute to the sky before it dives, the breaches they make when playing. They are the most social, the most dynamic and also the most communicative of all the great whales.

Their communication takes many forms, the most obvious of course being their song. In 1977, NASA cast the *Voyager* spacecrafts out into space, in a bold and beautiful gesture Carl Sagan called 'a bottle cast into a cosmic ocean'. Aboard were recordings of humpback whale song, and they are now drifting off into interstellar space beyond our solar system.

Acoustic expert biologists break whale songs down into motifs, phrases and riffs, and have recorded the genesis of certain 'tunes' in Eastern Australia. Within weeks the same song will be heard in islands thousands of miles across the Pacific, until finally they're being sung by unrelated groups of whales 9,000 miles away. It's like a catchy tune being hummed until every human on the planet has the earworm.

This is just one piece of evidence for cultural evolution in cetaceans, which also hints at broader implications of what whale societies might be. Traditionally we see humans as engaging in conversation when they're stood right next to each other, and we assume the same of animals. However, blue whales have been shown to communicate using infrasound across entire oceans. Who's to say in cetacean society those whales aren't conspiratorially chatting to one another, despite the vast distances between them?

At 9,000-plus miles the migrations of the humpback are the longest undertaken by any mammal, and one of the longest of any animal on our planet (only outdone by some seabirds, such as the Arctic tern and sooty shearwater). Female humpbacks are significantly bigger than the males, and they need to be. They mate, have their calves, and then nurse and nurture them in warm tropical waters near the equator, where there is little or no food for the adults. The calves continually drink milk that is actively squirted into their mouths, perhaps up to 50 gallons a day. This enables them to put on tonnes of weight, which they will need in order to begin their big migration to the colder, more nourishing waters of the Arctic or Antarctica. The females can go five months without food, just living off their blubber stores, which have to be built up in phenomenal amounts during the summer feeding season.

A decade ago I was witness to a so-called 'heat run' in Hawaii, as half a dozen males chased after a beleaguered female, vying for the right to mate with her. First of all the female will splash at the surface, her pectorals flaying around making sounds that carry for miles. This 'come and get me' advertisement attracts the males, and the chase is on. We swam along underwater as the whale train went by, males trumpeting like elephants and lurching their heads above the waves. Occasionally they'd slap each other with their tails or the caudal peduncle (a nodule on the lower part of the spine). It looks like a play slap, but the power creates whirlpools under the water, and can badly injure or even kill other suitors.

Breaching is not unique to humpbacks, but no species does it quite as regularly or spectacularly. After one leapt no more than five metres from our boat in British Columbia – so close that the spray soaked us – we watched the footage back in slow motion. What you don't notice in real time is how much they twist as they propel their gargantuan bulk clear of the water, fins swinging free like maypole ribbons. And they certainly seem easy with close human contact; out kayaking in the twilight off northern Vancouver Island, I had a leviathan the size of a submarine surface silently alongside me, so close I could have run my hand down her black gleaming flank, the bushy spray from the spout soaking me in what smelled like damp, fishy cabbage.

When they're engaged in feeding they can easily lose touch with smaller objects in the water, as I discovered when watching them bubble-net feeding off the southern coast of Alaska. Camerawoman Justine Evans and I undertook a stunning week-long sea kayak expedition to get the perfect shot of this event, the most elaborate and dramatic of all feeding spectacles. As many as 15 whales coalesce, swimming in from the horizon to one gathering point, then the whole group dive beneath a shoal of herring. One whale circles the fish, blowing out bubbles from its blowholes, creating a shimmering curtain that the fish will not swim through. Looking down from a helicopter or drone, the ring of bubbles appear like a circular snail shell spiral. The fish are trapped within, and will not swim out through the bubbles. Once the fish are trapped, the other whales let rip with a squealing screech to disorientate them. Finally, all the other animals erupt to the surface with their mouths open, the 26 pleats of their throats billowing, taking in thousands of gallons of water and tonnes of fish.

Justine and I had been filming the spectacle for around five days in the stunning inlets. There was never a minute of any day when we didn't have humpbacks around us, completely oblivious to our presence in our tiny two-person kayak. Justine sat at the front with a high-speed camera; I sat in the back and powered the boat. Even though we had both seen the spectacle many times before, it was intriguing that you could never predict where the animals would come up. Often we saw flocks of seabirds flying towards what we assumed was the herring and we would all turn our cameras that way, but the humpbacks never ever emerged where we expected!

In the afternoon of the fifth day, the conditions were beyond perfect. The sea was surreally flat and calm, and we continually had whales around us, bubble-net feeding over and over and over again. As we watched, a group of whales approached from the south and another from the north. We sat still, waiting to see what would happen. Everything went silent. There was an almost otherworldly calm. And then the surface around us started to froth. I immediately put in my paddles and paddled backwards as fast and furiously as I could. And then almost in slow motion, 12 whales, each of which might have weighed 36 tonnes, erupted above the surface. They towered above us, six or seven metres above our kayak, their gullets bulging with food. Giant pectoral flippers (the largest appendage in nature at as much as six metres long) flailed around us. Despite having been in the midst of the greatest feeding force on our planet, our kayak barely wobbled, and as the huge beasts slipped back into the silken waters, they left little more than a mirrored glassy patch, known as a foot-print, on the surface.

The team and I pulled up that evening at the lighthouse at Point Retreat, not far from the Alaskan capital Juneau, with towering snow-capped peaks covered with Sitka spruce and tumbling waterfalls. The sunset took hours, but by midnight it was finally dark, with stars so close it was like you could reach out and touch them. We all decided to sleep out on the dock (for the first and only time in all my many months in Alaska, the skies were clear enough, and weather report good enough that we trusted we could sleep outside without a tent!) and we were sung to sleep by the mournful melody of humpback males, wailing their plaintive song to the stars. 'Impossible!' one marine biologist told me. 'The males only sing in their breeding grounds!' 'Impossible!' another marine biologist told me. 'You cannot hear the songs above the waves, unless you are in a boat and the sound transmits through the metal.' And yet every one of us heard the songs. They went on all night long.

On another trip, this time in Drake Bay off southern Costa Rica in the early noughties, a male humpback was 'escorting' a female, seeking to separate her from her young calf in order to mate with her. He came underneath our boat and then surfaced, lifting our prow out of the water in a moment that was probably as terrifying to him as to us.

My most emotive encounters though were in 2022 in Tahiti in French Polynesia, in the absurdly clear blue waters where the mothers come to nurture and nurse their calves.* The adults mate here, then head 4,000 miles south to the waters of Antarctica to feed through the plenty of the austral summer. Then the whales return, the heavily pregnant females to give birth to their babies.

The very young calves – those that were no more than weeks old – were strictly guarded by their mothers. We watched in respectful awe as a mother deep below us took her young calf under her body and beneath her pectoral flipper, nuzzling it below her chin. It was exactly as you would expect to see a penguin taking their chick under their wing.

A calf is not as adept at regulating its buoyancy as its mother, so it would position itself under its mother's chin so that her bulk kept it down at a safe depth. The female was inevitably slumbering

---

* That's calf singular. Whales are uniparous, meaning they have just one baby at a time. Twins occur in fewer than 1 per cent of pregnancies.

at around 15 metres, in some cases even with her eyes tight shut. And while the mother would sleep for 20 minutes at a time before surfacing, taking three big breaths and then diving back down to sleep again, the calf had to surface every five minutes or so.

Once back in position, the calves would nuzzle in to the mammary slits that sit far back down the body either side of the genitals and nudge the mother to begin nursing. The milk is 50 per cent fat with the consistency of cottage cheese, five times richer than human breast milk, and rather than the baby sucking, the whales actively squirt it into the mouths of the young. A female can express 680 litres or 150 gallons a day for the entire period of nursing, while never feeding herself – sometimes for as much as five months. She may lose 15 tonnes of weight, the ultimate self-sacrifice for her progeny. It was a deeply affecting and emotional moment.

These waters are essential for the life cycle of the whales for several reasons. Firstly, their main predator the orca rarely shows itself here. The warm surface waters lead to an 'island oasis effect', with protected shallows keeping a constant balmy temperature. This means the young do not have to expend calories in keeping warm and can instead spend their energy laying down the blubber reserves they will need to undergo their trip back to Antarctica alongside their mothers.

For our entire two-week trip in Tahiti, cameraman Rob and I kept a respectful distance from the whales, hoping against hope for a playful calf. That moment hadn't arrived. In fact, we often felt that however respectful and careful we were, the whales were always aware of us being nearby, and would slowly, silently move away from us. Most days we never even put our wetsuits on, and by day ten had given up all hope of getting into the water with them. And then, on the very last morning, we came across a pair of whales just after sunrise, no more than a kilometre from the harbour. On its first sojourn back to breathe, the calf saw our boat and almost cocked its head to one side in curiosity. It turned and swam up to us, pausing before turning slightly, circling around us in a big loop, then returning to the sanctity of its mother's chin.

Rob and I lowered ourselves into the water as gingerly as if it were a scalding hot bath, doing anything to avoid a sound or splash that might spook them, but we needn't have worried. The mum would wake every 20 minutes and almost imperceptibly rise up to the surface like a nuclear submarine. She had a perfect circular hole taken out of her left pectoral fin that you could look clean through to the water beyond (so perfect that it must surely have been caused by the bite of a cookiecutter shark).* Normally humpbacks are identified by the speckles and dapples on the undersides of their tail flukes, but this was such a noticeable sign that I'll be able to follow this whale's progress, possibly for the rest of my days. Each whale has a very high fidelity to a particular migration route, so she'll be back here for sure.

She would take three breaths, then drop down to her resting position. From the boat, you'd see a glossy 'footprint' left behind as if oil had been poured on the water. From below, with sunlight penetrating down into the blue, she'd leave above her a shimmering golden halo. No wonder so many ancient cultures have worshipped whales as deities.

Every time it surfaced, the calf got bolder and bolder. At one point it swam till it was nose to nose with me, as if wanting to kiss, before swivelling to survey me with a giant, friendly eye. The encounter lasted three long hours, with bashful curiosity becoming active playfulness. After a while, the calf span and

---

* As mentioned before, cookiecutters are sneaky parasites, taking quick nibbles from their prey, often at night, and it is not uncommon to see the wounds from their unmistakable bites on seals, dolphins, tuna and sharks.

twirled in front of me, urging me to come and dance. I swam in, but with caution. It might just be a calf, but humpback whale calves weigh a tonne, even at birth. This one would have weighed as much as a truck and was definitely more clumsy, and less aware of its power than its mother.

On one occasion the calf swam straight at Rob but then seemed to realise at the last moment that it had mistimed it, and slammed on the brakes when practically on top of him. A big tail flapped right on Rob's head, thankfully without the force to injure him.

A whale has never seemed more familiar to me than in those moments the calf looked back over its shoulder as if to say, 'Come on! Come and play!' Sure, when it returned to its mother and you got the sense of her scale, you were reminded that she was an aquatic giant as physically different to us as any animal on earth. But when she took the youngster protectively beneath her flipper, hugging it tight to her, when she nudged it away as if to say, 'I'm having a nap now, go play with your new friends,' and when he cavorted around us like a toddler ... suddenly we were one. Both air-breathing mammals that nourish their young with milk, and who share a common ancestor. Sentient beings that love and chat, and that serenade and court their partners. For the entire three hours, the songs of distant males echoed in our ears, and rays of sunlight speckled the animals' dappled sides. It was the most emotional and perfect interspecies play date I've ever had.

The mature male humpbacks also undertake the same epic odyssey to return to the coral coves of French Polynesia, but they come here with one thing on their minds. The eerie tones of their courtship serenades are always in your ears if you dive beneath the surface. Sometimes you'll see a mother and calf pair with a male escort butting in between them, and on one occasion things got even feistier. A male and a female started to engage in exquisite breeding behaviour before us. They were oblivious to me and Rob as we hung spellbound and motionless in the water. At first they went through a range of intricate and affectionate embraces. They twirled about each other, with their mighty pectorals twirling like flamenco dancers' outspread arms. They came together nose to nose as if making to kiss, and they both arose in perfect symmetry, spyhopping out of the water alongside each other in a ritualised dance. And then from nowhere the female clearly decided she'd had enough. She turned her back on the male and swiped at him with her tail, driving him deeper below us. It was as if he had said something to offend her. It was the equivalent of watching lions mate: the courtship lasts for hours, but sometimes the female grows tired of all the attention and lashes out in a moment of unbridled fury.

Much of the youngsters' early lives are spent in building up their bodies for the travails ahead, testing their skills and tautening their muscles. In 2019, in the seas off the coast of Ecuador, we watched humpback calves breaching completely clear of the water over and over again for more than an hour at a time. Breaching in whales is usually described as being down to a desire to remove parasites such as whale lice and barnacles. Certainly one sperm whale we watched breach repeatedly was, as we discovered when we swam up to her underwater, visibly covered in loose skin, grotesque whale lice, barnacles and other parasites. Rather than the folded but smooth grey skin you expect to see on sperm whales, she looked like some kind of kid's art project, a whale macrame collage peppered with bits of repurposed garbage. This particular female looked skinny and unwell, and flying solo only had breaching to fall back on.

Breaching can also be used to transmit sound and communicate over vast distances, particularly via pec slaps and tail slaps. However, when calves breach repeatedly like this, the hypothesis is that they are in training. For an animal that can weigh as much as eight tonnes to get clear of the water takes a phenomenal amount of effort. To do it over and over again might be down to youthful exuberance, but it is perhaps more likely that they are building up the muscles that they will need for the vast and dangerous migration south.

Although orcas are scarce in these warm waters, once the humpbacks migrate into the temperate and then polar seas, orcas are abundant and will target their calves, driving them underwater and not allowing them to surface, trying to separate them from their mothers. These harrowing chases can last for days as the orcas seek to exhaust the calves. When they do finally tire out the youngsters, they might only eat their tongues.

Humpbacks also have to undergo the natural annoyance of other animals seeing them as playthings. In the far north I have seen Steller sea lions leaping, jumping and diving all over humpback whales. The humpbacks trumpet their breaths and swing their tails, clearly infuriated by the excessive attention. They often have escorts of dolphins, and can carry a coterie of parasitic fish like remora. They are in effect a giant floating ecosystem. In the crystal-clear waters of Tahiti, you can tell when a whale has been by, as the water is filled with tiny rainbow particulate – fragments of the skin they are constantly shedding, as well as gobbets of whale milk and the odd bit of whale excrement.

• • •

For hundreds of thousands of years, predators have been the greatest threat that the humpbacks have faced on their migrations, but now they need to run the gauntlet of our modern, busy seas.

The humpback is an animal for which we have universal affection, and yet there are so few structures that govern how we protect the seas for them. In French Polynesia we kept our distance and respected the animals in the water but had to endure seeing cavalcades of jet skis and speedboats thundering through their waters within metres of the animals as they slept and guarded their youngsters. While there is genuine risk of the calves being struck (one washed up dead in the lagoon just days before we arrived), there is another more universal but even more insidious danger. Noise.

One aspect of cetacean behaviour that science is racing to understand is the world of sound a whale inhabits. Sperm whales are possibly more dependent on acoustics and more sensitive to sound than any other animal on earth. They generate concentrated bursts of biosonar that bounce back off objects in the environment, before being received and processed as vibrations through the whale's jawbone then transmitted to the inner ear.

To get a crude sense of how biosonar works, close your eyes, make a clicking sound with your tongue and get a pal to wave their hand in front of your face. You'll be able to tell when their hand is there, due to the deader, closer sound of the echo. Some visually impaired people have trained this capacity themselves to do this to such an extent that they can resolve fine detail and manage to find their way around.*

Sperm whales use an organ called the phonic or monkey lips, smacking them together much as we do with our lips. These click sounds head out from the snout at 3,555mph and are probably processed or intensified by passing through the spermaceti organ. They can not only build up a 3D picture of their world based on the reverberations, they can also communicate over vast distances, using a vocabulary of sounds of extraordinary complexity.

---

* I was once lucky enough to work with Juan Ruiz, blind since birth and yet able to cycle around me in circles on a mountain bike, clicking his tongue against the roof of his mouth and hearing the responses.

To be underwater with a sperm whale is a complete sensory experience. In the waters off Dominica in the Caribbean it is possible to swim right alongside resting and socialising sperm whales, though it takes a good deal of luck and time to find your perfect animal.

I visited Dominica in 2011 to see and hear this for myself. Once out in blue water off the coast, our guide Andrew Armour engaged a directional hydrophone to get a sense of where the whales might be. This might sound like cutting-edge tech, but it is actually a scaffolding pole with a black rubber dish at one end and a microphone inside it. Twisting the pole around on board, you can get a sense of where sounds are coming from. When they are at their loudest, you're pointing in the right direction.

The sounds vary from chitter-chatter called coda that the females use to communicate at the surface through to the echo-location calls at depth. These consist of a number of clicks, which become closer together and quieter as the animal approaches its food (we assume – no one has ever seen this happen), turning into a grasshopper-like buzz as the sperm whale uses constant clicks to increase the resolution of their image. Thirty clicks per second provides a rapid update in the last seconds. It's absolutely identical to what you hear in a bat foraging for moths.

We didn't encounter any mature males in Dominica, but we did manage to record them hunting in very deep ocean canyons off Kaikoura in New Zealand in 2010. The biggest males emit some sounds at depth that have to be heard to be believed. These blasts sound like someone whacking a hammer on an iron pipe, and might well be the loudest sounds in nature, cranking up to 230 decibels.

Just as whalers once did in the bad old days, we spied our targets using their distinctive blow. Sperm whales have a single nostril skewed to the left, which gives them a unique spout that juts out at an angle rather than going straight up. You can tell it's a sperm whale from at least two miles away.

When you spot one, the temptation is to put the power down on the boat and get to the target as quickly as possible, but this is unwise. Speedboats hammering towards them make the whales nervous. In most cases they'll just dive. Even if you don't spook them, after a while you start to get a sense for animals that don't want you around. They start moving, becoming less inquisitive and less willing to put up with divers in the water. In a situation like that, there is no point in even trying to approach them. They are faster than you could ever be, and all you'll catch is a glimpse of their disappearing tail flukes. What you want is an animal who is as intrigued with you as you are with it.

In Dominica, it was day five before we found ours. We'd been encountering a group maybe eight animals strong, all females who cruised around thick as thieves. They had with them one tiny calf, about the size of a Friesian cow, which was no more than days old. And that's where things got interesting. We had no way of knowing which cow was her mother. The females were collectively taking care of the little one, presumably as mum dived to feed. One particular nanny was a very distinctive female called Can Opener, as she had a unique semi-circle taken out of the flukes of her tail. This was almost certainly the work of the nefarious cookiecutter shark too. She appeared to be the calf's main carer, and Can Opener was a player.* It started with her changing direction every time we entered the water, heading towards us rather than away from us. Within a few hours, that had evolved into her swimming lengths alongside me, rolling over onto her side so she could check me out with one massive eye, figuring me out in the same way I was her.

---

* She has been encountered here for 17 years and researchers know her for her 'amicable personality'.

After another hour or so, things took an extraordinary turn. As we slipped into the sea, we would see Can Opener swim ahead of the calf then detour, changing the youngster's trajectory with her body, steering it towards us in the water. She was introducing it to its first ever humans, as if to say, 'Here's something new and interesting in your world, you want to learn as much about them as you possibly can.'

Can Opener's intimacy also stepped up a notch. As I swam at the surface, she would fall in alongside me, matching me fin stroke for fluke flap. As I dived, she would dive too, and as I span around on my axis or somersaulted underwater, she would do the exact same movement at the same time. She was so engaged with copying me that she nearly took out my cameraman Johnny with one mighty sweep of her tail. Mirroring another animal is almost always a mating behaviour, but there was no way this 20-tonne grey wrinkly giant was flirting with me … was there? No, far more likely she just wanted to make a connection across species in the only way she knew.

In many ways, male and female sperm whales are like totally different species. The sexual dimorphism is pronounced, with the huge males living solitary lives and being able to hunt right up into the Arctic and Antarctic due to their vast size. One intriguing side effect of this is that the male and female brains are comparable in size, despite the males being up to three times larger.

If we return to the encephalisation quotient, we humans are top of the tree, with bottlenose dolphins and our close cousins the chimps and bonobos coming in a close second. Sperm whales have brains that can weigh as much as ten kilos, making them the biggest of all mammals (slightly bigger than orcas); however, the boys and girls are not equal. The EQ of the smaller females is near double that of the males. One hypothesis for a large EQ is that big brains evolve to deal with sociality. It could be that the females developed superior intelligence because they live in large social groups where social communications and 'friendships' need to be cultivated.

The coda communication of females has been described as sounding like 'Sellotape being pulled off a roll'. This constant chatter of clicks and clucks mostly goes on between females at the surface and is thought to be social chitter chatter. The booms and clangs that come out of a big male when they're at depth are something else entirely. Having been in the water with sperm whales when they're properly letting rip, I can tell you it is overwhelming and deafening. It is also highly directional. The precise second a whale turns the barrel of its remarkably thin nose *directly* at you, the sound courses through you like you've just sat on an electric cow fence. It's like someone shining a powerful torch beam at you. You only get dazzled when it's pointing right at your face. It's the same when you're gazing down the barrel of a sperm whale. You can feel its 'sound rays' resonating through your whole body, even from a comparatively small female sperm whale that is just checking you out and not trying to do you any harm. The males may be four times the weight and produce sounds that are triple the volume.

This phenomenal acoustic ability could have many applications and has led to one of the most intriguing and exciting hypotheses in nature. The description of this superpower dates back to the 1980s, and it is known as either the acoustic prey debilitation hypothesis or the Biological Big Bang theory. The idea is that hunting sperm whales use blasts of sound to incapacitate prey in the water, discombobulating, stunning or even killing their targets with sheer blasts of sound. As military forces around the world experiment with sound as a form of both torture and non-lethal weaponry, there's a tantalising chance that nature got there first.

Recently scientists from Woods Hole Oceanographic Institute claimed to have debunked this theory. They recorded a range of sperm whale sounds, then played them at fish and squid in a pool. Even those subjected to the greatest volume of sound didn't die, so Biological Big Bang was declared debunked. While I respect the scientific method, in this case I have to question whether this is really solid evidence. Our teams have spent weeks listening to sperm whales on hydrophones, and the range of sounds is extraordinary. And once you're underwater and in the firing line, even the smallest sounds can throw you off your game if aimed straight at you. Who knows what concentrated blasts can do?

There are several recorded instances of humans experiencing the effects of directed sperm whale sound cannons. Andre, who was one of our safety divers in Dominica, described to me an incident when he accidentally found himself in between a huge bull and the female he was trying to court. He tried to explain to me the sensation of being hit by that sound.

'I cannot even describe the pain,' Andre said to me. 'My whole body turned to jelly. I couldn't move. It was fierce – the worst pain I have ever felt.'

'How did you manage to survive?' I asked.

'Well, he was just giving me a warning; just letting me know who was boss.'

'And this was a sound, right? Could you hear it?'

'Yeah, man.' He nods. 'And when I got out the water my ears were bleeding ... but the pain, I *felt* it ... not just in my ears, but everywhere. Through my whole body, like a bolt of lightning.'

And this was just a short zap, intended as a warning rather than to maim or injure. The loudest sounds, those clangs of the males down at depth, are deafening even from miles away. No one has ever recorded let alone experienced one of those up close and directed. I am yet to be convinced, therefore, that they do not use sounds as part of their hunting repertoire, especially if the aim is to overwhelm their prey's senses.

. . .

None of this shouting and whispering is any use without hearing. Whales might not have external ears, but they have all the necessary internal ear apparatus, and in many cases sounds are transmitted to the ears through their jaws.* Orcas can tell apart the different species of salmon from the resonance of air in their swim bladders. Sperm whales can hear the difference between fish and flotsam, blue whales can hear their compatriots across entire oceans and humpbacks communicate over vast distances by slapping their pectoral fins on the surface.

The most frightening experience I've ever had underwater was diving under a fishing boat that was catching squid at night. There is no blackness quite as unsettling as the sea at night, and nothing so discombobulating as having thousands of pint-sized market squid bounced between your hands like little luminous bath toys. The clonking of the boat propeller seemed to be right on top of us, despite the fact that it was 30 metres above. And then the whole world exploded with a bang so loud it seemed to be inside my head, followed by a pressure wave that reverberated through every air space in my body. Disorientated and deafened, I would have made a lethal dash for the surface if cameraman Simon hadn't grabbed a hold of me. The fishermen above were throwing fish bombs, consisting of half sticks of dynamite, into the water to scare sea lions out of the squid nets, and the properties of water meant the sound

---

* This can also work in humans – I have a set of underwater headphones for swimming that don't go anywhere near my ears; they just rest on my cheekbones. You can hear the music as if it is playing inside your head!

and pressure wave travelled at great speed and without dissipating in force. Another went off, and another; it felt like we were being carpet-bombed. It was terrifying. I have never felt such dread and claustrophobia. Ears ringing, heads throbbing, we had to get out of the water, and fast.

Seal bombing – where explosive devices are thrown into the water to deter seals and sea lions from fishing grounds – is just one of the ways we humans are filling the deep with our noise pollution. A hundred years ago the ocean had the crackling of shrimp and dolphin whistles as its soundtrack. Whales used infrasound and ultrasound to communicate with each other across oceans. But everything has changed. Most modern boats have depth finders and fish finders that send incessant pings off into the blue. City-sized cruise and cargo ships have vast propellers you can hear from miles away. Seismic surveys for oil, gas and undersea minerals create pulses of intense noise, pipelines and radio signals hum and buzz, and offshore construction, such as the building of wind farms and bridges, pounds the depths with pile driving, pumping and explosives. It's a marine nightclub playing non-stop happy hardcore FM, an aquatic construction site with a never-ending jackhammer baseline.

What must it be like for marine mammals who can't escape the water and who rely entirely on sound as their primary sense and have aural systems so sensitive they make our piddly ears seem utterly pathetic? Sound travels further and faster through water. And endless noise is a key cause of stress. Ever had a noisy neighbour? Or spent a night in a hotel with an amorous couple next door? Anyone who thinks a small thing can't make a difference should spend a night in a room with a mosquito. Special Forces recruits are played white noise as the most effective form of torture, and playing 'I Love You' from *Barney and Friends* to inmates at Guantanamo was as effective as waterboarding.

There is no escape for whales and dolphins. We turned our oceans into a nonstop cacophony of noise that befuddles marine mammals, and perhaps drives them to strand themselves on our beaches. These often ignored strandees have in recent years been joined by the ultimate cetacean celebrities – errant whales, dolphins and porpoises that have wandered up the Thames or the Seine and become cause célèbre for city dwellers, who are suddenly made aware of these exotic new visitors. However, the stories of these exciting vagrants very rarely have happy endings. Most often the disorientated animals end up dying, their stinking corpses shocking unsuspecting city folk when they wash up on shore. One thought as to why such adept and perceptive species end up getting quite so lost, is that they are overwhelmed and confused by white noise.

Strandings of sperm whales around the North Sea intensified in the late 1980s as acoustic soundings were being increased as part of the search for oil and gas. These depth charges crisscrossed the whole area and seemed to correlate with sperm whales changing their migration routes, heading from the safety of the deep ocean and into shallower waters, where they stranded. In 2002, twelve sperm whales and a fin whale stranded shortly after a seismic search for oil was taking place. The oil companies dragged the corpses out to sea and sank them with dynamite.

In 2016, the beaching of 29 sperm whales along the coasts of the southern North Sea (in Germany, England, the Netherlands, France and Denmark) again called attention to undersea soundscapes. The first hypothesis of why the whales became stranded was that they'd strayed from their preferred deeper waters into the much shallower North Sea and their biosonar had become confused. A comparatively shallow sandy bottom and loads of ambient noise from the busy shipping lanes caused them to become disorientated and to inadvertently swim into too shallow waters. Once they had beached on the high waters of a spring tide, there would be no chance of refloating them.

There is some evidence that mass strandings of whales have been happening for millennia. Cerro Ballena in the Atacama Desert is a site where layers of ancient whale fossils have been found, on what was a tidal flat millions of years ago. These strandings appear to have been en masse, and happened due to natural causes. However, in modern stranding events, scientists are eternally vigilant to anthropogenic causes.

In 2006, a science team were tracking vocalising beaked whales when a military submarine on manoeuvres came through. They heard the sub's acoustic pings on their hydrophones, then seconds later all the whales they had been tracking for days disappeared. It is perhaps no surprise then that active sonar has been implicated in numerous whale strandings, with a range of species beaching in large numbers following submarine military exercises. The US Navy has admitted its high intensity sonar caused the beaching of beaked and minke whales in the Bahamas, and 14 beaked whales that started to beach in the Canary Islands four hours after naval exercises were studied and found to have 'severe, diffuse congestion and haemorrhage, especially around the acoustic jaw fat, ears, brain, and kidneys. Gas bubble-associated lesions and fat embolism were observed in the vessels and parenchyma of vital organs.' The latter effects were hypothesised to be injuries from the whales surfacing too quickly. Essentially, they got what divers call 'the bends'.

Another study, by the Space and Naval Warfare Systems Centre in the US, linked 39 beaked whale mass stranding events in the modern era to naval activities, although it concluded 87 other events were not linked. Beaked whales and blue whales have been proven to change their behaviour, stop feeding and move rapidly away in response to simulated navy sonar.

Much of what scientists know of this phenomenon comes from studying the inner ear of necropsied whales, dolphins and porpoises that wash ashore. No bigger than a thumb, these delicate tissues are inside the cochlea and need to be removed precisely before they start terminal rotting.

Michel André, director of the applied bioacoustic lab at the Technical University of Catalonia, has spent decades studying the effects of man-made noise in the ocean. His research has shown that seismic surveys can have a range of effects, from damaging the fine sensory hair cells to causing lesions and what amounts to the destruction of the ears. André's team can recognise the specific frequencies that are causing the damage and whether it occurred within weeks or hours of death – therefore providing strong evidence that it is sound that caused a particular death.

'If these hair cells die due to over-exposure to noise, the cells are dead forever,' Michel said in a recent interview. 'It makes a scar which can form within minutes after exposure to the noise.'

They are finding that most of their subjects have been deafened by loud and probably artificial sounds.

Strandings could well end up being nothing more than misadventure – a bachelor pod getting lost, for example – with the social nature of the beasts causing them to follow their fellows into tragedy. What is certain is that the silent seas are quiet no longer, and in oceans filled with anthropogenic anthems, a whale's super senses could eventually be its undoing.

• • •

In 1990 on my wanders around Asia, I spent a week in the tiny village of Lamalera on the remote eastern island of Lembata, in the Indonesian archipelago. Lamalera is one of the only places in the world that has been granted protection as a subsistence whaling community, its practices having barely changed since the 1600s. Every day I rowed out to sea with the whalers, in their tiny wooden boats with sails of woven palm. When they sighted a marine creature, the harpoon thrower would launch himself from the prow of the boat onto its back with spear in hand. It was a brutal pursuit, where the whales definitely had the edge, and gave a real sense of the challenges faced by the very first whalers.

The people of Lamalera had a huge respect for their prey, gave thanks to the heavens every morning before setting sail, and blessed the dead body of their catch if they returned successful. Not a single part of the body was wasted; everything was used in the village, passed out equally to every family. They had an intimate connection also to the characters of the animals they hunted. The whale shark was *ikan hiu bodoh* – 'the stupid shark', for it failed to flee or fight their harpoons. Most respected was the sperm whale. They only managed to catch one every few years, as the animals were so canny and seemed not only able to outrun and outgun them, but also to outwit them. When they talked of *ikan paus* – 'the Pope fish' – it was with hushed tones and awe at their abilities and fearsome power.

Before the era of explosive harpoons and powered boats, the early European whalers had a similar knowledge and respect for their quarry. They kept detailed records of their catch, and reported on changes in the behaviour of their prey. These records have been historically analysed using modern computer modelling, and have shown some intriguing things about sperm whale social behaviour. Records showed that when whalers took up new fishing grounds, to kill sperm whales was almost embarrassingly easy. However, within the space of a year, their success rate dropped by at least 50 per cent. Scientists looked at the data and proposed several hypotheses: first, that those initial hunting forages took out all of the 'easy' individuals; and second, that animals who had evaded capture knew better the next time they encountered a whaling vessel.

However, it turned out that much more important was social learning, with the whole population figuring out how to evade capture and then passing on that knowledge to their peers. It was shown that before we started to hunt them on an industrial scale, the only natural predator that a sperm whale had was the orca. In response to an orca attack, sperm whales would line up in a static group, powerful tails all aligned for battering their foes, heads with potent sonar like cannons ready to fire. This was the worst possible strategy for dealing with a whaling boat, though, which was presented with the proverbial fish in a barrel, motionless and ready to be blasted. Within the first few months, sperm whales would learn that the technique for evading sailing whalers would be to swim away fast upwind, where the boats rowed or under sail could not follow.

Later, with more mechanised boats, the whales learned to dive as deep and long as possible to evade capture. Whales that had never encountered humans soon took on the same techniques, showing that the new knowledge was being passed on animal to animal. Whales were exhibiting language, learning and transmitted knowledge. Even so, we still managed to take at least 3 million animals of varied species from our seas in less than a century.

One of the reasons whales were hunted in such numbers was that the heads of many cetaceans – porpoise, for example – contain an oil that is a near perfect lubricant. It maintains its viscosity in both high and low temperatures and resists penetration by dust. The spermaceti of the sperm whale is not quite as fine, but occurs in extraordinary amounts in just one whale. A large male might have 2,000 litres. From the 1820s on, this oil was used to lubricate the cogs and wheels of clocks, firearms and industrial machinery.

The whalers' precise records included individual sizes, weights and numbers. So much of our modern knowledge of these animals sadly comes from these records. This includes knowledge of what they eat from studying their stomach contents. It turns out that sperm whales prey on good-sized sharks and large fish, but as much as 95 per cent of their diet is formed of squid. Hunters recorded how long they could dive for when hunting or evading capture, and timed them staying under for more than two hours. They also caught individuals who had just consumed bottom-living sharks in waters where the seabed was three kilometres in depth, giving the first indications of quite how far they plummet when hunting.

The International Whaling Commission was set up in 1946 to regulate this inevitable drive towards extinction, with the Antarctic whaling fleet behaving more like a mining operation than a fishery, with no attempt at sustainability. All commercial whaling was banned in 1986, and specifically in Antarctic waters since 1994. However, membership of the IWC remains voluntary, and their rules are only of value if nations stick to them. Norway, Iceland, the Faroe Islands, Greenland and Japan continue whaling, although Japan is the only of these nations that hunts in the waters of the Southern Ocean.*

Filming at the whaling station of Bunavoneader in the Outer Hebrides, we watched black-and-white footage from the glory days of the station, where whales were once dragged up a

---

* Somehow the most objectionable part of Japan's continuation of whaling at the bottom of the world is that they justify their catch on a nonsense loophole. They state their catch is for scientific research, yet no research of any benefit comes from their endeavours, and the meat from their catch ends up on the table in Japanese restaurants. In the modern scientific era, all the research Japan is claiming to do can be achieved through DNA analysis, tagging, tracking and acoustics, without ever harming a single whale. In the name of spurious research, more than 15,000 whales have been killed since the ban.

concrete slope to be rendered down in big vats. One of the chimneys was still standing, but most of the buildings were gone. The Norwegian firms that drove the industry essentially made the factories flat-packed, ready to be moved elsewhere when the whales were hunted out.

The saddest thing about visiting Bunavoneader was seeing records of the kinds of whales they took. From the late 1800s till the whaling station closed in the early 1950s, they were regularly catching blue whales in British seas – over 400 of them in the years Bunavoneader was operating. The whalers reported that fin, humpback, Sei and blue whales were abundant in Scottish seas in the early 1900s.

Even today the numbers have not recovered, and may never recover. The great whales take many years to reach maturity and to breed – a blue whale takes 31 years. Their numbers are steadily building back up but are still believed to be as little as 3 per cent of what they were before 1900.

Today I've still never met anyone who's seen a blue whale in British seas, although numbers of minke, harbour porpoise and dolphin appear to be slowly increasing. But is it too much to dream – if we were to create a giant marine park, perhaps somewhere rich and wild like the Sea of the Hebrides – that one day blue whales may again frequent our waters? Surely that is a goal worth making some sacrifices for?

CHAPTER TWELVE

Pod Life

There's no question dolphins are smarter
than humans as they play more.

Albert Einstein

*Amidst the tangled mangrove roots, the waters are stained with
reddish brown tannins leached from a billion trees. It feels like swimming in warm Coca Cola. A flurry of tiny silver fish shimmer, then
duck back into the intricate sanctity of the root systems. Something
is coming, an alien predator too bizarre for words. Squeaks and
creaks pierce the water, then a staccato clicking. A long thin beak,
lined with up to 136 identical needle-like teeth, then a pink bulbous
head. And how pink! Nothing prepares you for quite how pink a
pink river dolphin, or boto, really is. Their lurid colour tones are
blanched beneath the tannic waters, but when they crest the surface,
it's as if an inflatable flamingo toy has popped above the stream.*

Everything about the boto is odd. While most dolphins have fused neck vertebrae, the boto is mobile, enabling them to look over their shoulders to snap a fish that might be swimming alongside them. Also unique are the tiny, piggy, recessed eyes, almost useless in these turbid waters. And with their bald-looking heads reminiscent of Franciscan monks', their triangular backs and their stumpy flippers ... they look like a joke animal. But then they begin to hunt and exhibit their fierce side. Piranhas, wolf fish, candiru – the demon fish of the waterways are a snack for the boto, and a pod of them is one of the most efficient swimming machines on earth.

Despite their ferocity, the dolphins swim about us, snatching fish from my fingers.* They're wild animals, but astute enough to know which bit is Backshall appendage and what is fishy snack. Then abruptly, the food is finished, the feast is done and the dolphins are gone.

• • •

Since *On the Origin of Species* was published in 1859, conventional biology has held true that the animal that is most like us is our closest relative, the chimpanzee.† However, as marine biologists have found new ways of comprehending what lies beyond the studious eyes of dolphins beneath the waves, more and more are starting to ask an intriguing question: 'Is any animal on the planet more like a human than a dolphin?' They throw tantrums, scheme, can be creative and solve problems, show empathy, fear and love. In captivity dolphins have recognised themselves in mirrors and shown awareness of self. Analysis of their language has provided evidence that they have names for each other. They

---

* This encounter took place 15 years ago – I wouldn't hand-feed a wild cetacean now. The ethics of interactions with wild animals are evolving, and generally for the better.

† And its close cousin the bonobo, both of which share 98.8 per cent of their DNA with us, and are in the same animal family as *Homo sapiens*.

can count, use tools, and play tricks on each other and their human counterparts. They cheat at games, give presents to their friends and lovers, and even grieve their dead.

Twenty years ago, marine biologists would have turned up their noses at all I've written here, and accused me of over-sentimental anthropocentric interpretations of animal behaviour. Now scientific consensus is swinging towards accepting these truths to be self-evident.

The bottlenose dolphin has a bigger brain than us humans, although that is not a decisive factor in which species is more intelligent, as they are bigger and heavier than us. The big question has always been, why? It may seem from our human perspective that a big brain is best, but the most successful animals in the history of our planet have not had big brains.* Big brains require a huge amount of energy, and they are delicate and subject to a vast array of traumas. Animals do not have big brains unless they are somehow vital to their survival. Surely the most exciting thing about cetacean studies is quite how young this science is, quite how much we have left to learn. It feels like a whole field of biology that is in its infancy and will unquestionably yield discoveries that now seem like science fiction.

· · ·

I've shared thousands of encounters with dolphin species in every one of our oceans, from the Arctic to the Antarctic. I've had the largest, the orca, breaching alongside my kayak, drenching me in the spray from their blowholes before rolling to one side to eye

---

* There are many factors to consider in what makes a species successful. If it is purely numbers, then a myriad of invertebrate species leave us for dust. If it is time around on the planet then everything from the jellyfish to the dinosaurs outlasted the puny 6 million years we've been around. The only metric that we'd 'win' would be the species most capable of changing its world, denuding its own resources, and eradicating everything else around it.

me inquisitively. I've been dragged alongside a boat in Drake Bay, Costa Rica, (battling to keep my goggles and trunks on!) as a superpod of a thousand spinner dolphins leapt and spun on their axes in dazzling acrobatic displays. I've swum with tiny Hector's dolphins no bigger than Labradors and freedived at the mid-Atlantic ridge alongside scarred Risso's dolphins looking like weird white bananas. These encounters have given me unspeakable joy, and the certainty that there is nothing so enriching as genuinely interacting, even communicating with, an obviously sentient member of another animal species.

That sentience is amply illustrated in the hunting methods shown by dolphins in the Bahamas. The most playful and abundant species here is the spotted dolphin. They're extremely curious and interactive, but get bored very easily. The best way to maintain an encounter is to provide a novel stimulus for them. As with all animal encounters, everything has to be on their terms. Swim after them or into their personal space, and they'll disappear and you won't see them again. However, if you can do something intriguing while pretending to be totally disinterested in them, the next thing you know you can find yourselves dancing a pas de deux with an entire pod.

With this in mind, camera operator Duncan Brake mounted his camera to the front of an electric DPV (driver propulsion vehicle) and I took another myself. The first we knew of spotteds nearby was a cavalcade of coursing dorsal fins carving through the waves, making a beeline for our bows. Then the dolphins dropped onto the bow waves created by our prow as it ploughed through the sea, just like surfers riding a cresting barrel.

Bow wave riding is a conundrum to evolutionary biologists, who tend to think of everything in nature as having a purpose relating to the fitness of a species and providing it with an evolutionary advantage. If no such advantage existed, then why would such a behaviour persist, they ask. This doesn't work with bow wave riding. Some scientists have posited that it is simply an energy-efficient means of travel, but even limited observation shows this cannot be true.

In the freezing depths of the Southern Ocean in 2014, my team encountered Peale's and Commerson's dolphins who gave up on feeding for entire days, purely for the chance to spin around the clumsy human intruders in their world, showing off their submarining skills and clearly relishing the contact with other intelligent beings. Our interaction with them occurred in one smallish bay. We would drive our boat across it, and the dolphins would fall straight onto the bow waves. Then, when we got to the other side, those same dolphins would fall back onto the waves and ride with us back to where they'd started. They did this for hours without letting up, before we finally decided we should find another place to film or they'd never get anything done in their day!

If bow wave riding was genuinely to save energy, then those dolphins in the Southern Ocean had totally failed. And if the behaviour was for energy-saving purposes, why don't other migratory ocean animals do it? Could it be social, the scientists ask? Could they be training their muscles for travel at high speed? Is it possible that they are removing dead skin or parasites? None of these things rings true. Instead, what is instantly obvious to a lay person who has witnessed bow wave riding is that dolphins simply do it because they can. Just like surfers coursing down cresting point breaks, they do it for kicks.

• • •

Dolphins live in what's called fission–fusion societies. This means that their groupings are constantly in flux. I like to think it's a little bit like being a human teenager. You have breakfast with your family, squabbling with your siblings over your Coco Pops, before heading into school. Maybe you start the day with an assembly, coming together with all the pupils and teachers in the school. Then you go to class with, say, 20 of your peers. At break time, you hang out with a few pals (sometimes the same ones, but not always), before moving on to a different class. At the end of the school day, you socialise with a whole new bunch of friends, before heading home to your family again. Or maybe (if you're a kid like I was) you take yourself off on your own for a few hours and just go for a bit of a solo explore. A dolphin's day is much the same, moving between different social groupings and 'friendships' depending on what they're up to.

Dolphin behaviour can also be highly compartmentalised, with the whole group travelling together, playing and socialising, or of course hunting. These complex dynamics mean dolphin watching gets better the more time you spend with a group. To begin with you're just enchanted by their beauty and their acrobatics, but within days you start to notice patterns, relationships, abrupt changes in attitude that are probably triggered by language and can result in the entire pod being ignited into a different mode of behaviour.

Predation by the dolphin family is the most creative of any animal group. The bottlenose dolphins around Bimini have a unique and bizarre feeding strategy not shared by the more sociable spotteds (with whom they interact, and even undergo interspecies social relations with). To film this we needed to be underwater alongside them.

The Caribbean Sea might be absolute paradise for us humans, but there are ecosystems within it that are extremely challenging for feeding cetaceans. Just offshore of Bimini are seemingly endless sand flats in around ten metres of water. Swimming over the top of them, they appear to be a barren desert, devoid of life. But the bottlenose come here nonetheless. It may be that in these crystal-clear waters it is just too challenging to catch shoaling fish in broad daylight, so they need to improvise.

We picked up our first pod of dolphins on the very first hour of a ten-day trip, on which our sole goal was to see their remarkable feeding behaviour. The dolphins lazily rode with us from the very northern tip of the islands all the way to the very south. For most of the day, they seemed super sociable, messing around with each other, doing high-speed flybys past the boat, even joining up with neighbouring spotted dolphins, which they seemed to boss around. They have even been known to have sexual encounters with them, pinning the animals to the seabed.

We could see from the surface that the dolphins weren't hungry, so we sat in our scuba gear, hot and uncomfortable, with a gathering puddle of sweat forming in our wetsuit gussets and leaking out of the ankles to form big salty puddles on the deck. It was four o'clock in the afternoon before Nick – one of the Bahamian 'dolphin whisperers' – shouted out: 'They're feeding, they're feeding!' Camera op Duncan and I shifted to the back deck, hearts thumping, as our boat the *Gone Astray* roared into reverse, churning great walls of white spume behind us, threatening to sweep us off the dive deck and into the water.

'Neutral, neutral!' Nick yelled to the captain, then, 'Dive, dive, dive!' to us. We slipped off the back and into the warm waters, bubbles clearing to reveal aquamarine shades and flat sandy bottoms, with six dark dolphin shapes coursing across them in unison. Bottlenoses are big, and are the absolute boss underwater. The biggest males can be four metres in length, and can smash most sharks with ease. This means that they can be genuinely intimidating. Here, though, they were focused, intent on their prey. The only exception was one young calf, perhaps six months in age, who flipped 180 degrees and came to circle us, but then seemed to be summoned by its mother and returned to the formation.

The dolphins swam less than a metre from the seabed, noses tilted down so as almost to graze the silt as they swam. Normally being in the water with a group of bottlenoses is deafening. The constant clicks, creaks and whistles screech right through you, ranging from joyous to nails-down-the-blackboard intense. Here the dolphins were silent – or at least to my ageing ears. My kids might have been able to hear a fair bit, but most of the sounds made by the dolphins would have been beyond even their keen juvenile eardrums. These are the high-pitched clicks of echolocation, which penetrate right down through the seabed and can differentiate between buried stones, starfish, and the nutritious razorfish and sand eels that are buried there. When a dolphin senses something hiding in the substrate, it emits a characteristic blast known as the 'razor buzz', before making an abrupt right turn* and then plunging down into the sand with its rostrum or beak.

We were finning like crazy to keep up with the dolphins, who barely seemed to be moving their tail flukes but were cruising along at deceptive speed. Then they'd nose the dust, before plunging into the sand up to what would be their necks. On occasion, we saw them plunging in all the way up to their pectoral fins, doing a vertical headstand with their bodies standing up tall in the water column and their tails wobbling about all over the place to keep them buried head down in the sand. Then they'd inevitably come up smiling and chewing on a tasty razorfish.

In places where the dolphins have been most active, the bottom of the sea is a lunar landscape, covered in craters like the surface of the moon. Each one is the result of a successful bottlenose rummage, hence this behaviour is known as crater feeding.

---

* Scientists have shown bottlenose dolphins turn right 99.44 per cent of the time.

Ten years earlier, just across the narrow straits from Bimini in the shallow seagrass meadows of southern Florida, I was lucky enough to witness bottlenose dolphins performing a very different, dramatic feeding behaviour. My camera and sound team and I set out in a flat-bottomed aluminium boat, its small engines creating guttural chutt-chuttering noises as we searched around the shallows of the Florida Keys. Within an hour or so we picked up our first small pod of dolphins. There were about six animals, but we kept our distance, watching them through binoculars as they trawled around perhaps a hundred metres away from us. If we got too close, there was the possibility that we might be more interesting to them than feeding – we didn't want to put them off their stride.

Below us, the water was perhaps waist deep, with swaying fronds of streaky green seagrass covering the muddy bottom. Buried in it were billions of invertebrates, crabs, squid, cuttlefish, marine worms and anemones. However, the dolphins were not interested in this paltry bounty – they had bigger fish to fry. Our local marine biologist lifted her binoculars to her eyes. 'Move in a bit,' she suggested. 'It looks like they're targeting food – they won't pay any notice to us when they've found something.'

Now we could see our small pod start to move in unison towards land. The water got shallower and shallower as it approached the scrub and mangroves at the shore, and as the tide fell, the mud flats were being exposed. Soon the waters were so shallow that not only were the dolphins' dorsal fins exposed, so too were their whole backs. My camera operator Johnny cursed at us all. 'Keep still the lot of yous!' he chided, balanced precariously, trying to focus his long lens on a tripod on a bobbing boat. Even breathing too hard made his camera wobble all over the place.

The dolphins powered into the mud, in a melee of white water and spray, then from nowhere a silvery curtain of fish started leaping into the air, and the dolphins' smiling beaks popped fully out of the water, catching the slippery fish like Labradors with bouncing tennis balls. Lacking the ability to chew, they threw their heads back and swallowed their catch whole. Looking down from the sky, you could see that the lead dolphin swam in a decreasing circle, creating a spiralled snail shell of a wake. The mullet felt cornered by this 'mud ring' and leapt to escape it ... into the smiling jaws of death!

· · ·

Mud ringing, as it is known, was first witnessed in Florida in 1999 and has never been seen anywhere else. Fifty miles away on Bimini, crater feeding was first recorded in 2000, and again is totally unique to the area. These two populations of dolphins do not appear to be intermingling, or sharing their life hacks. In Shark Bay in Australia, female bottlenose dolphins have been seen collecting sponges and using them to protect their snouts as they feed among the coral, a tactic that has never been witnessed outside of Shark Bay. It seems one dolphin comes up with a novel behaviour, then their compatriots observe it, try it out, then give it a go themselves. The behaviour spreads throughout the community and soon all of the dolphins are doing it. Learning through observation is something that not long ago we thought only humans did. It's now known that many primates do it, but we are only starting to scratch the surface of cetacean learning.

There is undeniable evidence that dolphins teach their young how to hunt, something I've witnessed myself in macabre detail. On the shores of Patagonia, one of the most brutal and impossible of all hunting strategies takes place. At a time when the tide is just right, the conditions are perfect, and fur seals stray too close to the surf, fully grown orca (which can top the scales at nine tonnes) will thrash ashore, deliberately beaching themselves to then banana their bodies back and forth, and eventually pull their quarry back down into the waves. My team and I hunkered down for five days on the one beach where this has been documented, waiting for the tides to be exactly right and this greatest of feasts to occur. The beach stretched out to the horizon, with violent waves crashing on the shore and chilling Southern Ocean winds sandblasting the fur seals as they wandered along the water's edge. What they didn't know was that lurking out of their gaze was the world's finest predatory force.

People often comment that I must have supreme patience to do my job, sitting in hides or on stakeout for days on end, enduring night searches where we might take weeks to find a single snake species. The truth is that anyone in this game has a positive nature, which means they always genuinely feel like the next minute could be when the harpy eagle makes its appearance, or that the next log or rock turned over could be the one hiding the bushmaster we're looking for. We don't get bored, because it's always just about to happen. This was the first time in my career that this wasn't true.

We had spotters several miles to the north and to the south. The orcas' approach could be seen from miles away, meaning that we'd often hear at least an hour beforehand that they were on their way. Additionally, we knew that an attack could not happen at low or mid tide. However, we still needed to sit there on the beach so as not to spook the fur seals and disturb their natural behaviour. We didn't know if or when the orca would turn up, but we did know it wouldn't be in the next few hours. It was too windy to read, too wet to curl up in a sleeping bag, and we had to keep quiet so we couldn't talk or play games. I have never been so bored in my whole life (and I've read *War and Peace*).

It was the final day, and the tide was almost at its zenith – providing the orca with the optimum depth for attacking, but also sufficient incoming tide that they would not get properly beached – when we got our first call from the spotters that the pod was inbound. It was less than half an hour before their classic black dorsal fins were carving through the spume, thundering towards the attack channel with seemingly impossible speed. They didn't muck about.

A young seal pup and its mother were splashing about at the shore, seemingly oblivious to the danger lurking mere metres away. The pod of orca stopped dead, just floating in place with their dorsals cresting the waves close to shore. How did the seals not see them? Biologists who have placed hydrophone arrays in the water report that at this moment prior to the hunt the orca go into stealth mode, with all communication ceasing, instead operating on either a pre-ordained plan or some other in-the-moment communication we don't yet understand.

It was a mature female orca who led the charge, beating her tail with huge downward strikes, churning the water to foam and thundering up onto the sands. She had slightly mistimed it, however, so that the female seal was between the orca's mighty mouth and the seal pup. Using her own body as a shield, the heroic seal braced against the orca's snout, like a rugby player shoulder charging into a tackle. She lashed around with her teeth, striking with extraordinary bravery at the vastly larger foe, and in the ultimate David (Davina) vs Goliath contest she bested the mighty predator, who retreated thrashing into the swell.

Just minutes later, another pup was not so lucky. This one had clearly been in the sea just beyond the break and was unaware of the predators' presence. The orca surrounded the youngster, shepherding it into a pen of flashing teeth. What followed was not an easy thing to watch. The orca would take it in turns to slap the baby, tossing it into the air with their tail flukes. It would then compose itself, and make a break for freedom, only to be slapped again and again and again. The seal was tiny. They could have finished it off and swallowed it whole. But they didn't. They played with it for perhaps 15 minutes, until finally it succumbed. Even more disturbing, we found several fresh bodies of seals on the beach that bore the classic toothmarks of orca. They'd been killed, then abandoned, and not fed on at all.

In 2022, while filming orca up in the fjords of northern Norway, our crew watched in sickened fascination as a group of orca repeated this exact same process, but with a little auk, a bird that is no bigger than a blackbird. This hunt went on for even longer than the seal hunt. It was just like watching an hour-long game of beach volleyball, with the orca batting the ball up into the air over and again, showing off their skills. I don't believe that anyone who was watching could have doubted they were enjoying the sport of it. Why in evolutionary terms would a whole pod of orca spend an hour of their time and energy on a tiny bird that wouldn't sate even one of them? Even more strange, why, once the bird was finally dead, did they just leave it floating there?

To the average observer it appears to be nothing more than malicious cruelty. An evolutionary biologist, on the other hand, would go to the other extreme and say that there must be a 'purpose' for the behaviour, or it would not have persisted. Well, my own answer to why is illustrated by another orca encounter in 2012, when we had ringside seats off Vancouver Island as a pod of orca took turns to assault an adult male Steller sea lion.

The bull Stellers are the biggest of the sea lions and the most intimidating, at three metres in length and weighing (in some cases) more than a tonne. Underwater they are frankly terrifying and the apex predator here. However, the pod of orca isolated this most challenging of targets, driving him into a bay and blocking off his escape routes. Over the course of the next hour, the orca would slap the sea lion with their tails, throwing him bodily from the water, then breaching clear of the water themselves to land their nine-tonne bulks on top of him.

The disturbingly brutal process continued until all the fight had gone from the sea lion. At the surface, beaten and exhausted, he lay terrified and spent. And then the orca calves came in. Potentially a year or so old, they were in all ways miniature versions of their parents, and mirrored their attack strategies to perfection. They slapped the beaten seal with their tails, breached to land on top of him and spyhopped to watch his movements. Every now and then, an adult female would nip in to deliver a cautionary slap to the exhausted seal, in order to ensure he could not fight back. And then finally, when the teaching session was done, the orca didn't tuck into their meal. Instead, on an unseen command, they all turned and swam away. There was enough meat and blubber to feed the entire group, and they left it behind.*

So-called 'surplus killing' in orca has been well documented, and particularly along the eastern Pacific coast. And while at first it may appear to be wasteful and cruel, surely this is just another sign of the dolphin's tendency to compartmentalise things in their day? If they're travelling, they're travelling; if they're socialising, they're socialising; and if they're at school, then the animal they have targeted is a teaching aid and not food.

Nothing would surprise me about orca. They have arguably provided the richest seam of whole organism biological research in recent years, with understanding of their communication and evolution gathering pace. One of the most exciting discoveries is perhaps the revelation that the orca species contains a

---

* We assumed the Steller would die from his injuries, but he limped away to a safe corner of the bay and was seen several weeks later swimming around apparently fine – such is the resilience of these blubber-armoured behemoths!

host of different 'ecotypes', distinct groups of orca with different appearances, dialects, prey species and habits. Though their ranges overlap, these groups do not appear to interbreed or even socialise with other ecotypes, and may not have done for tens of thousands of years.

In the northeastern Pacific alone, there are resident orca, which are fish feeders, and Bigg's or transient orca, which mostly feed on mammal prey from seals and porpoises to whales. Both of these live in small groups and most individuals stay in their mothers' pods for life. There is also an offshore ecotype which is little known but appears in groups that may be up to 50 strong, and from their tooth wear probably focus on sharks and rays for food. The orca that I've swum alongside in the Atlantic are from totally different ecotypes and might turn out to be as distinct from these Pacific ones as different species. As always, the most exciting thing about this field of research is quite how much we have left to learn.

● ● ●

The more time you spend in the company of dolphins, the more you get the sense that they would be more fitting custodians of the planet than we are. They display intellect, camaraderie, empathy. Despite their tantrums and occasional cruelty, they really seem to embody so much of the features we regard as being the best of being human. It is therefore even more sickening to bear witness to the flagrant disregard shown to their kind. Particularly when it comes to the catching of orca for captivity and the inhuman cruelty that comes with it.

The most shocking examples of our wars against our cetacean cousins are true causes célèbres in conservation, particularly the annual slaughters of dolphins in Taiji, Japan, and in the Faroe Islands. Both of these hunts are continued with stubborn pride under the tenuous banner of tradition, and both nations see international pressure as unwelcome interference with their proud native cultures.

In the Faroe Islands, the annual *grindadráp* has been taking place for centuries. The main target is pilot whales, of which they take around 700 a year. The animals are herded from the open sea and driven into contained bays, where the Faroese butcher them with hooks and knives. The waters are stained bright red, and images of the panicked and terrified animals thrashing up on the shore or in the shallows are too graphic for many people to stomach. Local people see it as a social occasion, as sport, and as a part of their culture dating back to the ninth century.

The anti-whale hunting organisation Sea Shepherd report that around 265,000 cetaceans have been taken since records began in this method. Entire pods are killed, and as these are family groups, this means taking several generations at once.

The *grindadráp* hovers on the outskirts of the news, as it is too brutal to be shown on conventional media. However, in 2021 it hit international headlines when 1,428 Atlantic white-sided dolphins were driven ashore and killed in one day. This not only shocked the world, but people in the Faroes as well; particularly because rising levels of mercury and PCB in these top-level predators mean it's unwise to eat much of their meat. Similarly, in 2008 the chief medical officer of the Faroe Islands stated that pilot whale meat was effectively highly poisonous and should ideally not be consumed by anyone, but definitely not by children or pregnant women.

On the other side of the planet, in Taiji, a similar bloody scene takes place. In an event immortalised in the multi award-winning documentary *The Cove*, dolphins are either killed for meat or prime specimens taken alive for the domestic and global live animal trade.

Japan is a country I love and know well. When I left university in the mid-1990s, I went to live in Osaka, one of Japan's largest cities. I spent most of my year in the magnificent judo dojo in the centre of Osaka-Jō park near the castle. It was a grand old building with huge pillars running down the sides, and the dojo had one of the finest reputations in the nation. At summer training sessions, there could be a thousand black belts on the mat at any one time. The noise and clamour was visceral, even more so as I knew I was about to get pounded by a Japanese grand master with skill levels I could only dream of.

I would also train in the much more modern SeidoKaiKan karate dojo in town. Sweaty and serious, it was a training hub for some of the finest mixed martial artists from around the world, where you could spar with legends of the fight game such as the 'blue-eyed samurai' Andy Hug. I immersed myself in Japanese culture, including studying the language till I was chatting like a local, and came to have a reverence for it that I hold dear to this day. I've been back many times over the years, and travelled the length of the archipelago, kayaking out to see Steller's sea eagles among pack ice in the frozen north and paddling around the beaches of the subtropical south alongside Japanese macaques. The longer I've spent in the archipelago, the more I've begun to peel back the layers of courtesy and ritual that run deep through this fascinating country, although, if I'm honest, the more I learn about the place, the less I actually understand.

On more recent filming trips, whenever I have posted anything on social media about Japan, I have been dismayed at the responses from my Western following: 'How can you support those barbarians?'; 'Don't you know what those sadistic murderers are doing at Taiji? #TheCove'; 'Evil whale-eating monsters, they should be strung up!' It seems wrong to me to suggest that just going to Japan is the same as gorging yourself on whale steaks or supporting those that do. But worse, I believe this isolationist, us-and-them, even racist point of view is actually one of the most damaging factors in the battle for conservation in Japan.

On a long car journey through Japan's sugar-frosted snowy mountains, I had a very illuminating conversation about Taiji with one of my Japanese friends. Hiriyuki is himself a wildlife photographer and nature lover. He is extremely polite, almost deferential. It was a surprise, therefore, when I tentatively asked what he thought of Taiji, and he responded with outright indignation. He was, as many Japanese people are, outraged at unwelcome meddling by Westerners in Japanese affairs. 'How dare you?' he said. 'After all you torture and kill foxes, badgers and hares for entertainment, just for fun! You don't even eat the meat.'

'That's not me!' I argued. 'And foxes aren't endangered!'

'Well, I don't kill whales,' he responded, 'and minke whales are not endangered either.'*

'But cetaceans have developed personalities,' I argued, 'clear ideas of self and others, they mourn their dead, care for their young, cry if they're separated from them ...'

'So do pigs and cows,' he responded, 'and you kill those in their billions. And bullfighting is absolutely barbaric, and in America you hunt everything that moves!'

'But I'm not American! Or Spanish! And I hate those things as much as anyone!' I was now starting to get pretty indignant myself, feeling defensive and argumentative. How could he lump me in with the trigger-happy Americans?

'But you're not doing anything about that, are you?' he continued. 'You're coming to Japan, and calling us inhuman,

---

* This is true – they're listed as 'least concern' by the IUCN.

and telling us we have to save whales, because that's what you Western people have chosen to save, and you say we have to obey your rules!'

He went on to quote a variety of different scientific papers and various moratoriums on whaling, and was clearly well prepared. Hiriyuki had been through this discussion before, and it really irked him. And yet he is an animal lover, does not eat whale himself, and thinks it's wrong for dolphins to be kept in small enclosures in aquariums. Every gentle question I asked was rebutted angrily, until I stopped asking to avoid the unforgivable rudeness of being impolite to a host.

As our long drive went on, the debate became less about the morality of killing dolphins and more about national pride. The issue was important to him *because* heavy-handed foreigners had waded into Japan and judged them. And I had to see his point. How would we feel if foreigners were to start calling Brits barbaric and evil because of our Sunday roasts? We'd tell them to get stuffed.

For the rest of the trip it kept me awake at night. I brought the issue up with other Japanese friends and received similar responses. There was a disconnect here that needed to be addressed, and conventional methods of diplomacy clearly weren't working. Over a fortnight travelling through Japan, as I rediscovered all the ritualised politeness of Japanese society, it struck me that what was needed was a more considered, understanding approach.

So I decided to return to Japan and go to Taiji. I would travel on my own dollar, though some friends at Sea Shepherd connected me with some of their people on the ground to get me started. Old friend and camera operator Simon Enderby offered to come with me just for the cost of his airfare, so we could record everything we saw.

It was a strange feeling to land in Japan as an unwelcome guest. Conservationists have been deported and banned from the nation for getting involved in the political situation when it comes to whales and dolphins. Much as I wanted to find out all about Taiji myself, I also didn't want to burn my bridges and become embroiled in a diplomatic incident that might hamper me from doing my job or of coming back to a nation I love.

Taking the train out to Taiji was a journey through a winter wonderland, with vibrant blue skies, and snow dusting the buildings and Zen gardens outside them. The hunt takes place from September to February, and I was ill prepared for the bitter chill of the winter wind. I purchased a too-small army surplus jacket and borrowed a tea cosy of a woolly hat, and still had to wear all of the clothes from my suitcase underneath.

When I arrived at the station in Taiji, in a moment of strange gauche horror, the wall was decorated with a colourful mural celebrating the wonder of the whale. Simon and I wandered down the streets from the station to my guesthouse and had ramen noodles in a little izakaya restaurant with flapping dark blue curtains hanging down outside the door. The sea shimmered with golden light, and little fishing vessels put-putted in and out of the harbour. It seemed the most normal and charming of small Japanese towns.

Next morning a taxi was booked to pick me and Simon up at 4am so we could get to the harbour in time to see the dolphin boats head out to sea. Jetlag woke me up way earlier than that, so I was up in good time and breakfasted on a can of sickly sweet hot coffee from a vending machine in front of my guesthouse before getting into the taxi. The silent alleyways and illuminated shop fronts were cloudy with a pre-dawn mist. Sat waiting behind my yellow cab was a darkened police car. As the taxi driver switched on the ignition, the police car's headlights illuminated. I had suddenly wakened in a film noir! The police car followed us all the way to a hill that led down to the harbour. When we got out of the taxi and started walking down the hill, the police car followed 20 metres behind us at walking pace. My heart rate quickened. I'm a law-abiding citizen, and suddenly I was being followed by the police in a foreign land. I realised I hadn't contacted the British embassy or anyone else about being here.

At the harbour, we saw three girls, all wearing black Sea Shepherd hoodies, with the trademark skull and crossed tridents on the back. They were fellow Brits, happy to see compatriots, kind and welcoming. All had been here for a while and had also been on several previous 'tours'.

As the fishing boats set to leave, Simon took his camera from his bag and started to film. Immediately the doors opened on the police car, which was waiting nearby. Two young, tall police officers walked briskly over to intercept us. I greeted the first warmly in Japanese, though in truth I was shaking with nerves.

'Ohayo Gozaimasu, O genki desu ka?' Good morning, how is everything?

He greeted me back in perfect, unaccented English. 'Mr Backshall,' he said, 'would you please explain why you are here and what you are doing?'

How did he know who I was? Was I on some kind of watch-list? Stupidly, I hadn't prepared myself for this. What was I supposed to say? Was I technically allowed to be here doing this? I didn't even know. I stammered out an unconvincing response, feeling as guilty as if I had a kilo of heroin in my backpack.

'Would you mind coming with me to the police station so I can check your papers?' he asked. Simon and I looked at each other. Papers? What did that mean? Did we need a filming permit? We simply had no idea.

He took us both to the police station, which was what is known as a 'Poly-box', a small building analogous to the old Tardis-style police boxes that used to stand on the streets of British cities. They checked our passports, asked for permits that we did not have, and asked us questions about our intentions. I responded that our idea was to watch, record proceedings and make some homemade films on events in the cove, doing our best to record what happened in an even-handed, respectful and understanding fashion. My plan was to do these films in English, and then deliver a simpler message in my sketchy Japanese, which I would distribute on social media in Japan. My sincere hope was that by making the effort, it would show that I was not being judgemental. I also hoped to appeal to my friends in Japan to perhaps come and look for themselves.

The police were not placated. They wanted to know if we were members of Sea Shepherd. When we responded that we were not, they warned us that speaking ill about Japan to the outside world could result in censure and in us possibly not being allowed to return. Simon and I tried everything to be friendly and warm and as open as possible. They didn't engage. There was no point where they were ever rude or aggressive, but we could sense their animosity.

Eventually they let us go, although by then it was midday and the action was over for the day. It turned out, though, that the whalers had returned with empty boats. They hadn't managed to catch any dolphins that day. For the next few days we repeated the cycle, going out at 4am every day. We'd watch the boats go out and then gather on a wooded hilltop overlooking the cove.

On the last morning, just as we had decided to give up for the day, one of the observers from the Dolphin Project who had been watching the boats through powerful binoculars shouted out, 'It's on, it's on!'

Everyone leapt to their feet. A hunt was underway.

It was an hour or so of horror I will never forget. At first we saw the dolphins maybe a mile out at sea. It was a small pod of Risso's dolphins. They were noticeably bigger than most dolphin species and carried the classic scarring and light colouration that sets them apart. Most were adults, but there were several young calves as well. The armada of small and manoeuvrable metal boats were gathered in a loose line behind them, forming a barricade around the animals and banging and beating on long metal pipes strapped down into the water, confusing them with sound and driving them in from the open sea towards the natural enclosure of the cove. As they got them into the last part of the bay, the whalers put out a ring and net around the bay to stop the dolphins escaping. The net was then tightened, moving them into a smaller area still – they were now trapped, and it was all over for them.

In an attempt to block out prying eyes, the whalers had erected a blue tarpaulin canopy over the bay. They couldn't, however, shut off our view completely, and there were two small vantage points from where you could look down through the trees and see what was going on below. About 20 people, mostly from Ric O'Barry's Dolphin Project and from Sea Shepherd, waited in those two spaces with cameras poised, looking down on the bloodbath. We couldn't see the full ferocity and horror of what happened, but what we did see will stay with me forever.

The water ran red with the dolphins' blood all the way out to the rocks, and the seas were beaten to a pink froth from the slapping of the dolphins' tail flukes. Most of the final coups de grâce were hidden, but one hunter was careless and set about a Risso's dolphin with a club in clear sight of the onlookers above. Other men waded through the water wielding long-handled knives.

But as any maker of horror films knows, there is far more terror in what you don't see, hints of the unknown allowing your imagination to fill in the rest. The sounds were far more horrifying than the glimpses of the massacre. The men worked in near silence, apart from the odd grunt of physical effort that it took to restrain and slaughter these heavy, muscular, thrashing creatures. But the calves squealed in terror, while the noises from the adults were slightly deeper and louder, vocalising their pain and confusion. I will hear those sounds in my nightmares for the rest of my life.

Finally, with the kill over, the boats headed out of the cove and away with the catch, the dead dolphins strapped to the sides. Again, in an attempt to prevent our cameras from recording the carnage, they placed tarpaulins over the top of the dolphins. The winds, however, flipped up the tarpaulins to reveal the dead dolphin carcasses underneath.

In the afternoon we met with a local fisherman who had once been a dolphin hunter but had turned against his colleagues. We sneaked off from the viewing promontory, avoiding the police escort (which stayed with us for the entire trip), and climbed into the back of his tiny old beat-up car. He had a secret way, around the back of a FamilyMart convenience store past the bins, then through some woodland and over rocky headlands to get to a vantage point where you could look down onto the dolphin pens. The Rissos we had seen in the cove were all slaughtered for meat. However, not all of the dolphins at Taiji end up this way.

Dressed in muted colours, we crawled on our bellies to the lookout point. Our new friend assured us that if the police found us here he'd be arrested, and we'd get thrown out of the country. Below us was another bay filled with sea pens, and inside were dozens of mostly bottlenose dolphins, bound for the aquarium trade around the world. In some of the pens, young dolphins were being taught simple tricks like jumping or spyhopping for food rewards. The pens were small, and they were bunched tight together. In the few hours we watched, we saw and filmed two animals being dragged out of the pens, presumably having died from exhaustion, stress or injury. They were hauled out using meat hooks.

One of the most bizarre twists to the Taiji tale is that the town is portrayed to domestic tourists as being a place to celebrate cetaceans. Murals showing the wonder of the whale – such as the one that greeted us at the train station – are found in several places around town. And the main public-facing side of the Taiji hunt is the Taiji-Cho Whale Museum.

This museum is one of the strangest places I've ever been. It is simultaneously a celebration of whales and dolphins, and of Taiji and its whaling history. Here you can see whaling memorabilia dating back through the ages, alongside tanks where living species are kept in small, tight circles. Japanese tourists mill around with their kids, eating okonomiyaki and processed meat on sticks, taking photos of the false killer whale and Risso's dolphins swimming around in the tiny pens. Others paddle out onto the water in little kayaks to get closer still to the inmates.

There is a long tradition of whaling going back to the twelfth century in Taiji, but the dolphin connection began when the town set up the museum in 1969 and decided they needed some dolphins for their outdoor tanks. Over time, the process has been refined and developed. Though most dolphins are bound to be sold for meat, each animal is only worth around $500 as food. The big money comes from selling live dolphins to aquariums. A live animal sells for upwards of $15,000, significantly more if trained. Most go to the hundred or so aquariums and swim-with-dolphin experiences in Japan, but also increasingly to satisfy the market in China.

Perhaps the saddest sight was in one aquarium with a walk-through section, housing a very famous individual. Caught here in Taiji in 2014, Angel is an albino, a white spirit animal of impossible beauty, as if she has been carved from alabaster. She and her fellows are kept in a tiny chlorinated water tank. Her eyes were smeared and bleary from the chlorine. In this area of the aquarium, music is constantly playing, and there is a looped Japanese voiceover describing what's on display. I found the non-stop sounds annoying, and I was only there for 10 or 15 minutes. Angel and her friends, with their hyper-sensitive hearing, are subjected to this all day, every day of their lives. It was a sad, sad experience.

I attempted to provide an evenhanded film of what I saw at Taiji, taking into account culture and being respectful of my Japanese friends and their traditions, as extreme as they might sound. I spoke with several ex-whalers turned conservationists. They covered their faces and insisted on having their voices altered to avoid persecution from their neighbours and government. I also tried to get some of the whalers or someone pro-Taiji to speak on camera, but no one would.

Many of the Westerners that I met there are good people who really care about wildlife. They were respectful of Japanese people and just wanted to bring about change. However, there were also wide-eyed fiery zealots, who separated into factions and would listen to no one. They were firebrands of hatred but didn't seem to have any real impetus to *do* anything to bring about change. Indeed, some of them seemed to want the situation in Taiji all for themselves, to have it as their own personal project for their own glorification. This is a situation I see over and over in conservation and is one of our greatest challenges if we want to obtain consensus.

My firm belief is that we can only achieve tangible results when we work together with other nations, with respect and understanding. Some will ask, why bother? Most years the death toll hovers around 2,000 animals, and most species taken are not endangered. With entire ecosystems on the verge of collapse, why should such a hunt be on our radar? Are there not more important causes for the global conservation community to be lending their support to? But for me, this particular issue is important because of what it stands for.

One of the impossible questions in conservation is how to prioritise our efforts and attentions. Are certain kinds of animals more important than others? Clearly some animals can be more than a sum of their parts – 'umbrella' species that can attract attention and cash, which can then be used to preserve their whole habitat. Icons like tigers and pandas, for example, are protected by creating a sanctuary of the entire forest they live in, thus providing an umbrella of protection for everything else that resides there. However, pelagic creatures such as dolphins that can span entire oceans don't fall into that category.

Most conservationists would argue that it is ecosystems and their broader health that are important, not individual species, and on a cerebral level I concur. However, the lifelong animal lover in me does not. We all have to draw our arbitrary lines in the sand, and mine are based on a fuzzy amalgamation of my own interests, rarity, closeness of relationship to humans and, above all, on intelligence and sentience. I will cheerfully go and catch a scallop for my supper, because I know it is not a rarity, that our nearest common ancestor was perhaps half a billion years ago, and because it has relatively simple neural processes. The opposite is true of cetaceans. All are negatively affected by human behaviour, many are endangered, and several others are extinct or close to it. So much of their society and lives mirror our own, and for those reasons, their captivity and slaughter is utterly unbearable to me.

It seems that the strongest chance we have of bringing these practices to an end will be economics. Japanese people just don't eat that much dolphin or whale meat any more. Whale was very much seen as a cheap post-Second World War protein source, with Japan consuming 200,000 tonnes a year. Now that is down to 5,000 tonnes, with retailers trying to rebrand the meat as luxury, a delicacy or an essential part of Japanese cultural heritage.

Showing that these creatures are special, that we will not countenance their destruction merely for tradition's sake, is vital. Because if we cannot save our closest relatives, if we cannot engender respect and love for the most bewitching and intelligent animals, then what hope do the other 2 million species we share this miraculous planet with have?

I also know that this cause is a good one because I work in the media, and I am a pragmatist. The gory slaughter of cute dolphins will turn heads, will sicken people in a way that destruction of mangrove swamps will not. And small-scale problems can be addressed and can be solved. Bring an end to the hunts of Taiji and the Faroes, and every person who's posted on a social media site or signed an online petition about the hunts will feel empowered, will know they can make a difference. And empowered, educated people can change the world.

CHAPTER THIRTEEN

Uncharted Seas

The fair breeze blew, the white foam flew,
The furrow followed free;
We were the first that ever burst
Into that silent sea.

Samuel Taylor Coleridge, 'The Rime of the Ancient Mariner'

Currents and eddies swirl around the flanks of the sunken volcano, tossing me flotsam style along the coral battlements, through clouds of silver and black jacks billowing like butterflies. The titanium form of a Galapagos shark slides by, an entourage of pilot fish swimming at the nose of their meal ticket. My dive watch beeps at me. 'You're too deep,' it says. 'What are you thinking?' A dribble of water leaks into my full face mask and billows round the glass like smoke. I look like an abandoned astronaut, tether lost while on spacewalk and cast off into the firmament.

High above, the spectacle coalesces into view. A hundred sperm-like silhouettes wriggling in squadron, hammerheads stark against the glow of the sky high, high above. For mere seconds, we are witness to the grandest and rarest marine sight, a shiver of critically endangered scalloped hammerheads schooling, moving in unison, for now safe from the fishermen's hooks. But then on some unknown signal the group splits and bolts, the sharks powering over our heads faster than you would believe an animal could move underwater. In a fraction of a second they are gone, leaving us to contemplate what superior predator could have made them take flight, and whether they will seek us out instead ...

. . .

Cocos Island loomed into view after two days and nights steaming through empty Pacific blue. It was 350 miles of open ocean to the nearest land (Costa Rica) and several thousand miles of nothing till you reached Antarctica in the south. The small island stands alone, a dot in the blue. Isolated as it seems when viewed from the surface, with a decent mariner's chart you soon see that Cocos is just the tip of a sunken seamount, a volcano with its foothills a mile below the surface. It is the highest peak of a marine mountain range to rival the Himalayas in scale. Cocos is Everest, but there are a thousand such seamounts in the range, they just don't quite crest the waves.

Cocos Island was the inspiration for Isla Nublar in *Jurassic Park*, and certainly looks like a place where dinosaurs might roam. Its steep sides jut fortress-like from crashing oceanic waves, and stacks of jagged raptor teeth erupt from the swell. Beyond the intertidal zone, everything is dense tropical forest. Skyscraper-high waterfalls tumble off slopes straight into the sea, and giant straight-winged frigatebirds with red throat balloons circle around the peaks like pterodactyls.

But it is what lies beneath that draws biologists and divers to Cocos. The ridge of the sunken mountain range is a highway for long-distance ocean migrants who navigate through the Eastern Pacific using the seamounts as their waypoints, hide-aways and hangouts. As a protected marine sanctuary, Cocos is the last bastion of one of those marine nomads: the scalloped hammerhead shark.

Our vessel was the *Sharkwater*, named after the film of the same name, made by late conservationist and shark activist Rob Stewart, who was lost on a deep rebreather dive looking for criti-cally endangered sawfish. On board we had a team of the world's finest shark researchers, led by Randall Arauz, the Goldman prize-winning* marine biologist who has made a career – and a firebrand reputation – by repeatedly taking his own Costa Rican government to court for their environmental transgres-sions. Our mission was to catch and tag migratory sharks such as whale sharks, scalloped hammerheads, Galapagos, dusky, silky and tiger, and to monitor their movements between here, the mainland, the other sanctuaries of Malpelo, in Colombian waters, Clipperton Atoll and the Galapagos. Our goal was to demarcate the 'no man's land' thoroughfares in between these havens, to try to create protected 'shark corridors' and come up with a plan to stall the almost inevitable thundering towards their extinction.

Randall stood at the handrail looking towards Cocos, with his salt and pepper beard, and a bandana tied round his head. Though he had been here dozens of times before, his excitement about the opportunity we had to cement the science and save these species resounded through the whole boat.

'These waters are protected by law,' he said, 'but go a couple of miles in any direction and it's open season. You can catch a

---

* Very much the Oscars of conservation, honouring grassroots activists from around the world.

hammerhead on a rod and line, or drop a bomb over the side and take out a hundred. It's the Wild West out there. No one sees what goes on, no one cares.'

Our expedition to this lost world would last less than two weeks, but Randall's hope was that the information we garnered would change those rules, change the perception and force people to care. We had 20 tags to place on and in sharks. Some of these were cutting-edge satellite tags, which are stabbed into the dorsal fin and stick to it for six months to a year. Eventually they pop off and float to the surface, able to tell you everything about where that individual shark has been, including the depths they've gone down to, and when they've been hunting, cruising, or just hanging out. Many of these tags are sadly returned to Randall by a fisherman who has caught that shark. Other tags stop broadcasting in one spot, as an industrial fisherman has just stripped the fins from the shark, and dropped its body into the water alive, to sink to the bottom and die.

The cheaper tags we have to implant beneath the surface of the skin of the shark, a heart-stopping enterprise that involves catching the shark alive, giving it surgery on the deck of the boat and dropping it back into the water before it suffocates. These are acoustic tags, which will ping whenever the shark swims near to one of the submerged sonic stations scientists have in place across the Eastern Pacific.

My job would be to dive alongside Randall brandishing a Heath Robinson-esque piece of science tech, made out of a couple of pieces of plastic plumbing piping. The rig took the shape of a young steer's horns, with laser devices where the horn tips would be. Between them was a miniature camera. The lasers shone parallel beams out into the water, which I would need to dot onto a shark's flank. Using the known distance between the beams, we could then extrapolate and record the length of the shark. So we would be swimming together in shark-filled seas, Randall with a harpoon to stick into a shark fin and me firing laser beams at a shark's head.

. . .

Dusk, and dark clouds hung over Cocos. Torrential curtains of rain had been falling for three days without let-up. The waterfalls tumbling from the high points had become thundering cataracts, and a veil of water gushed from every gully and precipice. The near shore waters ran café au lait brown, and we'd had to send all our dives deep to get below the zero visibility.

Randall stood on the dive deck at the back of the boat, loops of line at his feet, and rigged hooks into giant rotting tuna heads. Weighted ones were sunk to the bottom for tiger sharks, whereas buoyant midwater bait was used for Galapagos sharks, silvertips, blacktips and scalloped hammerheads. The yellow buoy at the surface bobbed and then sank straight down. 'Tiger shark!' yelled Randall, giddy with excitement. He'd yet to tag one of these stripy swimming dustbins here in Cocos.* Line whipped out from the deck as the giant shark swam off with the bait. For a few minutes the team gently tried to reel her in, but then the line went slack. She'd taken the tuna head but not the hook. We went back to waiting.

Catching sharks on a hook and line – even for science – will never sit easy with me. I've given presentations worldwide to more than a quarter of a million people about the catastrophic impacts of shark fishing, and the dangers of overfishing to these fragile and overexploited wonders. Nowadays, you cannot do a shark dive without seeing individuals that have been disfigured

---

* Tiger sharks are determined predators, capable of biting clean through a turtle's shell. However, their prime role appears to be as a detritivore, clearing up dead and dying animals.

by fisherman's hooks and lines; many animals will still be carrying several in their torn jaws. I've spent weeks swimming after species like oceanic whitetips with a pair of pliers in my hands, risking everything to try to free them.* Seeing even one shark get hooked sickens me to my stomach. However, there is also no doubt that without shark science, many species will disappear in my lifetime.

Some species are already as good as gone. In the Gulf of Mexico, once common oceanic whitetips have lost 98 per cent of their number. Most of the species I work with are critically endangered. Some, such as the great hammerhead, could disappear with a single change in fishing restrictions. Others, such as the little-known Ganges and Borneo river sharks, have not been seen for decades and are probably already lost.

So this one shark suffering the temporary affliction of a barbless hook in order to provide information that could help save its entire species isn't ideal, but right now it is the only way. When it comes to the ethics of working with sharks, no standpoint is universally agreed upon. But it still makes me squirm every time I see a shark landed – even when it's done by people like Randall who are on a mission to make a difference.

It was getting dark now, with the rain unabating, and even Randall had taken shelter inside. Then there was a twitch on the line again. I strained my eyes against the gloom. Yes! 'SHARK!' I yelled, and six scientists dropped their drinks and scrabbled to the line, which thrashed and whipped like a stuck snake over the

---

* Removing hooks is extremely dangerous. Most sharks don't realise you're trying to help them and the pain causes them to lash out. I've had a nurse shark pull away at speed dragging monofilament line through my hands and cutting my palm to the bone. And on another memorable release I cut a huge dusky shark free of 30 metres of line and the colourful lures attached to it. As I pulled out the hook, she and four oceanic whitetips spun around and attacked the lures ... which were in my hands!

sodden, slippy decks. It was not the seabed line this time, but the midwater bait that'd been taken. It was quite a crude process from now on, with the science team letting our quarry have its head and swim off, and then reeling it in. The shark needed to be tired before we could consider getting it on deck.

In most instances, we'd bring sharks alongside before flipping them over onto their backs. This sends them into 'tonic immobility', where their electric super senses are so hyper-stimulated that they go into a torpor, as if they've been hypnotised. In this tranquil state, the biologists are free to work on the animals without them undergoing any stress or anaesthesia. In this case, though, we were operating on the animal, so it would have to come on board.

As it came close we could see it was a Galapagos shark, perhaps three metres in length. This was the perfect shark for us, as they are a highly migratory species that traverses the entire Eastern Pacific, but little is known about their breeding and pupping sites. As Randall slipped a noose around its tail, it thrashed, drenching the crew. Eventually it quieted, and we hauled it up on board. The deck hand placed a running hose into its mouth to keep the gills oxygenated and a wet towel over its eyes to calm it. His was the most precarious position. If the shark thrashed now, its teeth would rip through his arms.

What followed was as practised and polished as a Grand Prix pit stop. Everyone had their job and worked at high speed. Dr Alex Antoniou, owner of the *Sharkwater*, timed and coordinated proceedings. 'Two minutes and counting,' he shouted. 'Come on people, six minutes and she's going back in the water no matter what.' My tool and task were simple: a tape measure to record snout to vent and fin length.

Sexing a shark is easy – the males have paired external sexual organs called claspers, dangling below the cloaca like two wobbly white wieners. The tag is the tricky job. A neat incision was made on the belly, and the thumb-sized transmitter seated inside. Three stitches were whipped in with a steady hand. 'Five minutes!' Alex called out. The race was now on.

'Clear!' the shout went up as the last stitch was snipped.

'Biopsy for genetic sample?' Randall asked.

'Six minutes!'

'No time, let her go!'

The towel and hose were pulled out, the rope was slipped, and our shark rolled overboard with a splash. The instant it hit water the right way up it sparked into life and thrashed away, clearly none the worse for the experience. The team all cheered, slapped backs, bumped fists. Our shark had just become a beacon for its kind, a vanguard at the sharp end of conservation.

· · ·

Our intention had been to dive all day every day at a site that Jacques Cousteau named 'Halcyon', declaring it one of the finest dives on earth. There's nothing to see at the surface, but 30 metres down there are a collection of gullies, canyons and cleaning stations that bring in large marine animals and vast shoals of fish. Sadly, the bad weather we were experiencing was just the sting in the tail of a hurricane passing over Central America, creating swirling currents and howling winds. Halcyon was too exposed – thrashing waves made it near enough impossible to travel to the location and getting into the water distinctly unwise.

Manuelita island was our second choice, but our only dive site there was milky and murky with sediment. This made it a safety call, as Mark the cameraman had been down there on a good day and a tiger shark had swum at him with clear predatory intent and smashed his surface marker buoy. Being around big bold tigers in poor visibility wasn't an option. The south of Cocos was also out of bounds because of the crashing swell. This left just one dive site open to us, somewhat underwhelmingly named 'Dirty Rock'. This was an anonymous-looking boulder the size of a caravan, 'dirty' with guano from the brown boobies and noddy terns that perched there to rest. Although innocuous at the surface, it was actually a seamount summit, dropping down sharply to a hundred or so metres in depth. The currents carried nutrients from the deep steeply upwards until they coalesced in eddies around the shallows, drawing in huge amounts of marine life.

You never knew what you were going to get at Dirty Rock. On one safety stop,* a couple of giant sailfish hovered in front of us, resplendent in burnished silver, eyeing me up with their crests erected. We saw a dozen different species of sharks and giant schools of silvery jacks so big they blotted out the sun. There were always scalloped hammerheads, but they were surprisingly shy for such big sharks, and never came close to us. Occasionally you'd get a glimpse of a shoal of 20 or 30 moving together, but they were always too far away to be tagged. On one occasion five were cruising above us, then suddenly bolted over our heads with electric speed, faster than I've even seen a mako move.† Mere minutes later we heard the classic clicks and squeals of dolphins, and then a small pod of bottlenose cruised by. Sharks and dolphins are dire enemies, with sharks predating small species of dolphins and their calves, and dolphins known

---

* In diving a safety stop is usually a standard break of three minutes at five metres in depth to allow nitrogen to escape from the tissues. This is different to a decompression stop, where ascending from deeper and longer dives you might have precisely programmed stops that can last for many hours. Tech divers sometimes take books or waterproof iPads to pass the time. While you can in extremis miss a safety stop, missing a decompression stop can be fatal.
† The shortfin mako is the world's fastest shark, clocked at speeds of up to 50mph.

to deliberately flip sharks over to put them into tonic immobility, before killing (but not eating) them. Without question, these hammerheads were triggered by the acoustics of the dolphins and fled as fast as their tails would carry them.

The expedition ticked away. Entire days passed with us confined to our cabins as the rain lashed down on the boat. The $5,000 satellite tags sat on the crew room table as if mocking us. Randall would stroke them as he passed as if they were an amulet or lucky rabbit's foot. Eventually, it was the last day, and our final opportunity to dive. Despite constant rain, wind and high seas, we were desperate for some kind of magic conclusion, so we kitted up for a final attempt at Dirty Rock. The small team bounced out on our inflatable boats from the liveaboard* to the crusty white crest. Every wave threatened to ram my coccyx up through my throat and rattle the retinas from my eyeballs. Dive cameraman Mark was quieter than normal. I've worked with Mark for 17 years, and one of the things I like most about him is how much he lives and loves the job. He hates missing the shot even more than I do.

Boss Rosie counted us down from five, and we all rolled back over the side, driving down with our fins to get deep as quickly as possible. Turning my head even slightly to the left or right, the drag of the current threatened to yank the mask from my face. We saw the dark of the rock, and pulled into the eddy behind it that was forming a quiet swirl of slack water, filled with gargantuan shoals of fish, themselves taking shelter from the ferocious current. For all the ripping force of the water, the visibility was quite extraordinary (perhaps 50 metres) and the amount of fish life simply mind-boggling. This was surely going to be one of the greatest dives of my life.

---

* This refers to any boat that you sleep on board as well as working from.

Taking stock, we looked around. Our two safety divers were there, but no Mark. Where was he? I used the mic in my full face mask to call out for him: 'Diver Mark, diver Mark, please respond.' Nothing.

Then I spoke to the surface: 'What's happened to Markie?' No response.

We swam around for about ten minutes, looking for any sign of him. Nothing. Panic was starting to rise in my gullet. He was diving a rebreather, a system that recirculates the air you breathe and 'scrubs' carbon dioxide from it so you can keep breathing for hours. Rebreathers are alternatively known as 'bags of death', as if the system fails you get a fatal hit of carbon dioxide and asphyxiate. This was the same system that Rob Stewart was using when he was lost at sea.*

Our safety plan was clear: short search pattern then abort the dive, forgoing a safety stop. We dragged ourselves back onto the tiny Zodiac. There was no sign of Mark and grim faces on board. He had a new wife and baby son the same age as my twins. This was bad. My heart was in my boots. Never leave a buddy behind.

Then, bouncing over the waves towards us came the other Zodiac, and on board was a sheepish-looking Mark. His massive camera had acted like a sail, the current catching him and taking him half a mile out to sea. Luckily he had an emergency GPS beacon with him, which is enabled at the surface and had helped the crew to locate him. Regardless, it was a sobering moment.

Silent and sombre, we returned to the main boat and swapped out our scuba cylinders. When we returned to Dirty Rock, conditions seemed to have got even worse. The boat was

---

* These malfunctions, however, happen very rarely with modern rebreathers. The trade-off is that they give you longer time underwater, greater depth possibility and, crucially when working with animals, no bubbles, which many marine creatures hate.

tossed around like a shopping bag on the breeze, and the water looked black and utterly unappealing. Everyone sat as far away from Randall as possible – we were in the middle of a tropical cyclone, on an inflatable boat, and he was brandishing a six-foot arrow-tipped harpoon.

Rosie counted us down again. 'Five, four, three, two, one, GO!' We rolled and started the downward swim but were immediately stopped in our tracks. We appeared to have dropped straight in on top of *Thunderbird 2*, the big green spaceship. It was a gigantic female whale shark! As big as a school bus, she was a cruising fish-hangout. You rarely see a big shark without a few hangers-on. Remoras or shark sucker fish use their flattened slatted foreheads to adhere to a shark's raspy skin, getting a free ride and feeding off their scraps. Stripy pilot fish ride the pressure wave at a shark's pointy snout, driven along with relative ease by the larger animal. But this individual whale shark was remarkable – there were *hundreds* of big fish in a Tour de France-like peloton, hugging to the mighty form as it passed by with effortless sweeps of its tail.

Then, finally, we saw some scalloped hammerheads in the kind of shoals that draw people to cross the Pacific just to get here. As we dropped down to 30 metres and felt the bitter chill of the thermocline,* perhaps a hundred wriggling dots came together into a swarm above us. Each one had a dark blue hammerhead silhouette against the light sky.

The hammer-shaped head or cephalofoil is a wonder of evolution, and the many hypotheses for its function all probably have

---

* While water in the ocean is constantly mixing, layers of different temperature (thermoclines) and salinity (haloclines) are incredibly defined. You can usually discern them thanks to a strange swirly quality to the water. With a thermocline you can certainly feel it. Quite often when the temperature plunges, you'll almost instantly see deep water species such as giant sunfish or thresher sharks wandering up into shallower waters.

some truth to them. As I mentioned previously, all sharks have electrical sensors in their snouts.* Each is a narrow pore filled with jelly and linked to a nerve cell. They are so sensitive to weak electrical fields that they can detect a stingray buried in the sand by perceiving the charge created by the beating of its heart. This super sense also explains why these animals come to Cocos together. They migrate along the oceanic ridgelines by perceiving both the earth's magnetic field, but also more powerfully the magnetic iron particles in the volcanic rock. The hammershaped head spreads out these magical ampullae, making them more accurate. It also spreads out the eyes, giving the shark a broader field of vision. And finally the cephalofoil functions like a wing, giving the animal lift as it cruises through the deep.

As I contemplated the spectacle with awe, Randall came powering past me, spear in hand. I scrabbled for the laser measuring device and switched on the two beams. As soon as an animal came within reach, Randall struck like a gladiator, plunging his harpoon into its cartilaginous fin. The shark went off like a spooked chicken, but just minutes later was back, cruising past us in the blue. The tag hung perfectly from its dorsal. Bullseye! Hotshot Randall managed to look triumphant even through his mask.

In the months that followed we tracked our Dirty Rock girl more than 2,200 kilometres, as she headed to the Galapagos via the tiny Clipperton Atoll and Malpelo Island off the coast of Colombia, both of which form stepping stones that sharks use to traverse the Eastern Pacific. Another stopping-off point for our tagged hammerhead was a sanctuary for this and other sharks. Revillagigedo is a UNESCO World Heritage Site, and the biggest marine reserve and no-take zone in all of North

---

* See Chapter Six for an explanation of the ampullae of Lorenzini.

America at 57,000 square miles. It is more remote and better protected than any other site. It was to this most treasured of marine protected areas that my team were to journey next.

• • •

The *Quino el Guardian* chuntered across the velvet calm waters of the Sea of Cortez en route to Revillagigedo, pulling a steady if unspectacular six knots. I had the distinct sensation that I could swim faster. Unlike the *Sharkwater*, the *Guardian* didn't look like a cutting-edge space-age scientific research vessel that was pushing the boundaries of exploration in the Eastern Pacific. Because she wasn't. Instead she was all our budget could afford. The interiors were fake wood formica and the scuffed astroturf carpets were darkened by diesel fumes. I shared bunks with Nick the soundie and Mark the cameraman in a room the size of a broom cupboard. Poor Mark is six foot two, and could only sleep curled into the foetal position. But despite the rough-round-the-edges charm of our rust-bucket boat, we had unparalleled access to attempt exploratory diving on unmapped seamounts in one of the world's most prestigious marine sanctuaries.

My old friend Mauricio Hoyos (Doctor Shark to his friends) was taking Randall's place to continue the research of shark migration in the Eastern Pacific. Mauricio had tagged more than 300 sharks in Revillagigedo national park alone. Together with Dr Frida Lara, who completed her PhD on the movements of sharks in the archipelago, and Alejandro Gonzalez, the director of the national park, along with a handful of the world's finest marine biologists, divers and filmmakers, we had a formidable team. I had never been closer to Cousteau's iconic exploratory voyages on the ship *Calypso* in my career.

Cousteau referred to the Sea of Cortez as the aquarium of the world. It's almost impossible to imagine how it was in his day, before the factory fishing ships pillaged the waters with impunity. The modern-day story of the Sea of Cortez has been one of continual conservation crisis. The vaquita, the world's smallest and cutest cetacean (a porpoise no bigger than a cocker spaniel), has often been caught as bycatch by fishermen looking for the absurdly high-value fish, the totoaba, itself also critically endangered. The totoaba is targeted for its swim bladder, which is dried and sent to China to be used as a medicine. The pursuit of this useless organ has reduced the vaquita to no more than a dozen or so animals. It will almost certainly go extinct in the next few years.

That said, Cortez is still one of the world's great marine hotspots, with huge numbers of sperm, fin, humpback, grey and blue whales. It is also home to California sea lions, who breed at the rocks of Los Islotes. The superyachts you find here in La Paz, and increasingly in every decent harbour around the world, are a sign of one of the greatest threats to our planet. The rise of the super rich is something that you don't have to be a card-carrying socialist to find deeply disturbing. Some of the 150-metre-long superyachts we saw here cost upwards of a quarter of a billion pounds to buy. They have their own helicopters, subs, on-board yachts and other toys. Many of our crews have worked on them and tell tales of sickening opulence. Of each one being restocked each week with tens of thousands of dollars' worth of beluga caviar from Russia, bluefin tuna sashimi from the South China Sea, foie gras from Paris. At the end of the week, the consumables are all dumped over the side – the owners never come to the yacht at all. The boats are just kept waiting, stocked like a six-star hotel for every whim of a clientele that never checks in. Each one of these oligarchs, sheiks or tech billionaires may have half a dozen of these yachts from Monaco to Melbourne. The resources that are squandered by the mega wealthy put our own personal sacrifices and lifestyle choices into stark relief, and all feed back into our great challenge; to balance the books, and restore equilibrium to our beleaguered planet.

As the *Guardian* left the waters of the Sea of Cortez behind, it headed out into the Eastern Pacific bound for our destination. Revillagigedo sits astride two tectonic plates, and at the intersection of two ocean currents: from the north, the cold California current, and from the east the warm North Equatorial Current. Our expedition here had two purposes. The first was to dive and explore uncharted waters around the archipelago to attempt to uncover their secrets. The second was to tag and track as many pelagic sharks as possible – 40 per cent of the shark species here are threatened with extinction. To connect with the information from our expedition on the *Sharkwater*, some would be fitted with satellite tags, which would track their progress for up to a year, monitoring their course, how deep they dive and the temperature of the water they visit. Tagging a pregnant female could then give vital information as to animals' potential pupping sites and nurseries, which then become the most important places for conservation protection.

For the first 48 hours of our journey out to Revillagigedo the seas were high and heavy. Most of the crew lay out on the deck wrapped up in towels, faces drained of blood and going various shades of green and yellow. During the third night, though, the seas dropped, and we woke to a gently undulating quilt of blue. Flying fish startled by our prow leapt from the water and using wing-like pectoral fins powered themselves over the surface, sometimes going a hundred metres before they sploshed back in. Bushy sprays from humpback whales and the slick backs of blue whales also provided good cause to be up on deck gazing out to the horizon. By mid-morning the water was so calm it was almost eerie, bringing to mind an ethereal molten metal, deep blue mercury spangling with sun sparkles. Tiny black storm petrels fluttered around the surface with their bat-like flight pattern, tapping on the water with their feet to bring plankton up to feed on. They appear to walk on water like Saint Peter (hence the petrel moniker).

The closer we got to the island, the more we were strafed by masked boobies, gannet-like birds with stiff wings and torpedo-shaped bodies, built for harpooning into the water to catch fish. A pod of six bottlenose dolphins spotted us from the horizon, the classic falcate dorsal fin slicing as they leapt, and convened at our bows, dropping onto the pressure wave formed at the prow. They surfed along in front of us for the best part of an hour, the water so clear and so blue beneath that you could see every scar down their grizzled grey flanks.

We came to sight of land on the evening of our fourth day at sea. The grey silhouette was San Benedicto Island, the second largest in the Revillagigedo island chain, yet little more than a speck among all the blue. Just a mile or so away from the island, the seabed was still the abyssal plain at around 3,700 metres below us – vast, barren, unknown. Then the seabed soared upwards towards the island in abrupt flanks, to crest the waves in a barren, intimidating volcanic summit. San Benedicto was last active in 1953, in an eruption that lasted more than a year and seared the landscape free of any living thing. Now there is a large exposed caldera of ash with rivulets running down the sides, and black rock lava fingers stretching east low out to sea. There is still no vegetation on the island, but albatrosses and masked and red-footed boobies now make their nests in the slumbering crater.

We pulled into a crescent of low-lying black lava fingers forming a protected bay with the volcano as its backdrop. Almost immediately, silky sharks started nosing around our back deck. Silkies are named for their particularly glossy skin, which seems to have been finished in buffed titanium. They also appear to be the species of shark that is most driven to investigate sounds and vibrations in the water. When filming them in Cuba, we didn't use bait to attract them, but took old plastic water bottles down with us. Crunching the bottles up in our fingers and wringing them like a damp tea towel created vibrations in the water that brought the silkies nuzzling around us like bloodhounds on the scent. Every time they circled around us, they'd zip up to the surface and touch our hydrophone with the tip of their snouts, clearly intrigued by the weak electrical fields it was giving off.

Next morning at San Benedicto, we dropped in at the dive site known as Canyon, reputedly the finest dive site in the entire North American continent. The team was buzzing with excitement. We only had this one day to explore, but with conditions beyond perfect, we had a good chance of doing four long dives here and getting some sensational results for the science team. Mauricio's tags could add to the data we had garnered in Cocos. Just as there, our hit list started with a large pregnant whale shark and moved on to Galapagos and hammerheads. We were also adding to the list one of my personal favourites: the stunning silvertip shark.

The second I hit the water, I could tell it was going to be a special dive. The site, at around 24 metres below the surface, was a busy cleaning station, with barberfish, Clarion angelfish and thread-like wrasse in yellow and black longitudinal stripes. I had never seen so many silvertip sharks in my life, perhaps 20 of them. Silvertips are classic requiem sharks: sleek, handsome, extremely hydrodynamic, and with clean silver or white lines running down the dorsal and pectoral fins. If the silky is inquisitive, the silvertip is pugnacious. They swim directly at you, before banking away at the last second – an approach profile that in other species would be a warning of territorial aggression but in silvertips just appears to be a manifestation of their forthright nature. They are a large and slow-growing species, being born at around 68 centimetres and exceeding three metres when they're mature. Here, most were youngsters, not much more than pups. It turned out Canyon was a silvertip nursery, and further out in the big blue was where the adults congregated.

Silvertips mostly hang out in reef habitats and are less of a target for the longline fisheries that are decimating many of their cousins, but that doesn't mean we are not negatively affecting them. As a top predator, silvertips feast on other fast, open-ocean predatory fish. But that position at the top of the food pyramid has a flaw. Along with many other top marine predators, some silvertips contain in their tissues dangerous amounts of poison-ous heavy metals, such as mercury. As youngsters, silvertips mostly feed on benthic (bottom-dwelling) prey like skates, rays and octopus, but as they get bigger they'll feast on other pred-ators, like scad, jacks, trevally, tuna, wahoo and other sharks. These in turn have been feeding on smaller fish that might have themselves fed on plankton, and that plankton bioaccumulates poisons like mercury. Every step of the chain results in a higher concentration of these lethal pollutants. Mauricio's research has shown that even though we are 720 kilometres from the Mexican coastline, the silvertips are still laden with the industrial pollutants we flush out to sea.

Though the spectacle before me was spellbinding, it was clear from the second we settled on the bottom that something was wrong. I could see Mark the cameraman and Dr Frida Lara, who was my dive buddy, but no Mauricio. Then there was a crackle on the dive comms: 'Steve, Steve, this is Scott. Mauricio has a problem with his ears, he can't get down to you guys.'

I looked around me at the melee of sharks in this most beautiful of dive sites, and looked at my computer. I was at 27 metres. If I stayed here for the full dive we'd need a three-hour break before we could get in the water again, and we needed to get the tagging done.

'Call it,' I said. 'Let's abort the dive.'

We swam back up to the surface together, muttering and grumpy, but all knowing an ear squeeze is no fun, and that there was no point us continuing without our chief scientist.

Just an hour and a half later, the second dive started far better. We descended as quickly as possible down to the cleaning station. The visibility was off the charts. A small shoal of huge scalloped hammerheads (probably all pregnant females) circled around the site, joined by scores of other requiem sharks. I pressed the clearing device on my full face mask up into my nose to push out the pressure that was pressing on the airspaces in my ears. It consists of two round black plastic balls that seal up your nostrils, allowing you to equalise. For the first time in a thousand dives with the mask, it didn't quite seal. I looked down and one of the plastic balls fell off and rolled around in the bottom of my mask. My heart sank into my dive boots. Without that nose piece I couldn't equalise. If I went any deeper – even a couple of metres – I would burst my eardrums. I had no option but to abort the dive.

On the way up, I had to stop to decompress, and the bobbing up and down with the swell put so much squeeze on my ears that I had to rip the whole mask off and ascend without it. I shared air and borrowed a mask off my support diver Frida, but was completely attached to her, and dependent on her to get to the surface and safety. As I bobbed there in the blue, a giant oceanic manta flew around and around us, flapping its immense wings, going over my head to blot out the sun. Cameraman Mark was 20 metres below me and didn't see it.

By now, everyone was starting to get itchy feet. When we got in the water for the third dive, there were more equipment failures. Then the stills photographer's ears went. Abort. Fourth and final dive, and in a slight panic to try to get something at any cost, the team was doubled to eight people. Complicating things never ends well – underwater you keep it simple at all costs. True to form, we got split up in the current and didn't even find the dive site. Dive aborted. Zero sharks tagged from four catastrophic attempts.

It was a brooding and frustrated crew then that festered in our bouncing bunks as we headed overnight to our next destination. Roca Partida is a lonely seamount a hundred nautical miles east or west of any other small islands and 750 nautical miles away from the Mexican mainland. All you see as you arrive is a two-pronged rock like a manta's horned head. Front on, it's the size of a semi-detached home, but as you motor around it, you realise it's actually only a sliver wide. The top is stained white with millennia of bird droppings, and frigate birds, boobies and noddy terns perch on its steep sides. Some of the masked boobies even have chicks here, and regurgitate food for them as they huddle on their meagre ledges.

As much as it provides an oasis for ocean birds, beneath the waves it is an even greater haven for marine life and a visually arresting sight. The west wall is a vertical cliff face as steep and uniform as a skyscraper, dropping away into the abyss with such abruptness that it gives you vertigo. The east drop-off is also steep, but more protected from the prevailing currents, with balconies packed with resting sharks, fluttering pufferfish and lobsters the size of Jack Russells. At the north and south points, you have to fin with all your might just to stay still, while a show of epic magnitude plays out in front of you. Shoals of black jacks and tuna swirl about in the ripping flow, smaller fish feasting on the banquet created by upwelling nutrients, driven into the shallows by the predators.

This sole pinnacle soars up from the depths, blocking the Northern Equatorial and California currents, two mid-ocean rivers that have their confluence here. It creates the perfect storm. When we were there a shiver* of silky sharks a hundred strong hung off the point, wriggling sperms silhouetted against the sun. Nervous scalloped hammerheads kept their distance.

Because I had to thrash so hard to keep my best seat in the house, I took a nasty CO_2 hit[†] and left the water with a headache so intense I could barely open my eyes. We only had one more dive here, so I forced myself back in again. Two giant whale sharks floated past. One steamed straight over the top of me, oblivious to my presence. It was this dozy demeanour that led Indonesians to call it *ikan hiu bodoh,* 'the stupid shark'.

Mauricio, though, didn't miss a beat. He drew back his arm for a second, poised like Poseidon himself, before plunging our satellite tag into the dorsal fin of the giant beast. It didn't so much as flinch.

Divers here inevitably focus on the grand spectacle of the multitudes out in the blue – the migrants and nomads stopping off here on their epic journeys, using it like a motorway service station for meeting, feeding and breeding. However, some of the finest shark spectacles are to be found snoozing in the eddy of the rock. Slumbering whitetip reef sharks lie stacked like sardines in the caves, some with their pectoral fins over their fellows as if having a little cuddle. Open ocean sharks practise 'ram ventilation'; that is, swimming constantly to drive water through the gills and transfer oxygen through fine filaments into their blood. Whitetip reef sharks, on the other hand, use the buccal

---

* The collective noun for sharks.
† When you've breathed in an excess of carbon dioxide. Extremely dangerous.

pump in their throats to power water over the gills while resting motionless on the bottom. They laze here all day long before heading out at night to feed among the cracks and crevices of the reefs. Some of the females I saw were distended with pups, which they give birth to right here in litters of up to five. No more than metres away, hidden in fissures in the volcanic rock, were miniature versions of their mothers the length and breadth of my forearm.

One of the heavily pregnant mums bore brutal-looking toothmarks down her side. On the surface I relayed this to Mauricio. His response was to take out his laptop and show me footage he had recorded here at Roca Partida. He'd been swimming down to pick up one of his receivers when he'd happened upon whitetips mating. It was the most riveting, and unsettling, shark footage I'd ever seen. One target animal was chased and swarmed by around 20 others.

'That's the female, and they bite her by the fins and turn her upside down,' Mauricio described.

'It looks like they're trying to kill her!' I replied, and it did. They bit her around the pectoral fins and threw her around as if she was prey.

'The skin of the female is thicker so it can protect the internal organs. There are ten sharks mating with the same female. With some female sharks it's super difficult to tag them – I need two scalpels to cut into them cos of the thickness of the skin.'

At one point, a shark swam off with her in his mouth like a pitbull with a chew toy. She appeared dead. Five others followed. They were all fighting to insert one of their claspers into her cloaca.

'I've never seen – in any animal species – a mating so brutal.'

'That's the way it is – with sharks, love hurts!'

At this point, the two sharks sank to the bottom with their bellies up. They lay there lifeless, the female still clamped in the male's jaws. The others swarmed over them like cockroaches over a cake. It was a deeply disturbing image.

'It's not just sperm competition, is it? They can have two fathers.'

'Not just two …'

'One litter, many fathers?'

'And that's what's going on here, all these males battling not to be *the* one, but to be one of the fathers! And at the end, she looks like she's dead. But I saw her later and she was OK.'

This is surely as dynamic and dramatic as fieldwork gets. Scientists like Mauricio are inventing their methods as they go, and any one single tag can bring back information that revolutionises our perception of an entire species, even one we might think we know well.

Our ten-metre-long pregnant female whale shark was to be a true emissary for her species. Our state-of-the-art transmitter followed our girl for a year, before finally popping off and transmitting its recordings via satellite to Mau. During that whole time, she never dived deeper than 200 metres, feeding on the abundance of plankton and small fish near the surface. But then she suddenly and abruptly dived down to a mile in depth and stayed there for 24 hours, barely moving, before coming back up into relatively shallow waters. Mauricio believes she took this one dive to give birth to her pups at depth.* This is something we might never see happening. Until someone pushing the boundaries of biology can figure out a way – perhaps a long-running 'follow-me' drone – of tracking

---

* Little is yet known about whale shark reproduction, but the species is thought to have the largest litter size of all sharks. In the 1990s, researchers were able to ultrasound a pregnant female and discovered that she was carrying an astonishing 300 pups. It's estimated that pups are approximately 60 centimetres long at birth, though sightings of specimens this small are incredibly rare.

one of these animals with vision, then their private lives will remain a mystery.

Whale sharks are the biggest fish in the world. They are an animal icon, as recognisable as a panda or a chimpanzee. Yet we know next to nothing about their lives. We have never witnessed even the most basic elements of their life history, like mating and giving birth, and can only guess at them. The same behaviours in terrestrial animals have been seen and recorded since Stone Age times. It remains both the most frustrating and the most exciting aspect of shark biology. There is still so much left to learn and understand, and that hunger to discover will drive on the next few generations of marine biologists.

• • •

The next goal of our expedition was exploration of the undersea features around the distant island of Clarion. The furthest west island of the archipelago, it is another 24 hours steaming from Roca Partida. It only has three known dive sites, and nothing to rival Canyon or Socorro, so is generally ignored by dive live-aboards. However, with Alejandro on our team, we had been given unique permission to explore the undived sections of the coastline, and even to use DPVs (diver propulsion vehicles or scooters) in our exploration.

Originally we had planned on travelling an extra 30 hours out to a totally undived seamount called Hurricane Banks on the very edge of the marine protected area, but a severe weather warning had meant that we'd had to abandon that plan. Our research documents therefore included little information about Clarion, describing it simply as 'a low-lying extinct volcano'. Smarting over the loss of the totally new seamount, and without any glowing reports about Clarion, my expectations were low. They have never been so completely smashed.

As Clarion finally appeared as a smear on the horizon, bottlenose dolphins fell in at our bow. For the next four days, there was never a moment in the day that those on board didn't see a dolphin or humpback whale, and usually both. Masked boobies, elegant white tropicbirds, their flimsy tail streamers flouncing behind them in the sea breeze, rode the air above our crow's nest just as the dolphins surfed below. And the island itself was stunning – a mountainous volcanic crag, with one of the most impressive coastlines I've ever seen, made up of great black fingers and wave-carved arches, precipitous dark cliffs, and stacked wafer-thin sedimentary beds between the ancient lava. It was breathtaking and ... big! So much bigger than I had been expecting. I had thought that we might only have hundreds of undived metres to explore, but in reality there was going to be miles and miles of potential exploration. Suddenly, we were all Cousteau again, with untold miracles before us.

But just as we were revelling in these new discoveries and the potential knowledge they might provide, we were confronted with a taste of the challenges threatening this Eden. Sat just a mile offshore was the sport fishing boat the *Royal Polaris*. She was fishing for tuna with impunity, right there inside the marine protected area. High behind the boat were two kites, taking her hooks out behind the stern in order to give as many chances on the sport fish as possible.

This boat had been warned a dozen times in the past for similar offences within the no-take zone, but they just kept coming back. For the entire time we were there it anchored in the same sheltered bay as us, and spent its days pulling out critically endangered fish from the national park waters. It was one of many boats that openly advertise online sport fishing safaris into Revillagigedo. This is the ocean equivalent of someone brazenly advertising on Facebook, offering trips to go and shoot a mountain gorilla in the Rwenzori or a giant panda in Chengdu.

The fact is, the high seas are a wild frontier, and the temptation to break the law is great. Pacific bluefin tuna are the most valuable wild food on earth. The most expensive Pacific bluefin ever sold went for $3.1 million. And here is one of the only places that I've seen true giant bluefin in the last decade. The rarer the fish become and the harder they are to find, the more their price goes through the roof and the more sport fishermen will covet their illegal capture. And it's not just small-scale rod and line fishing. Mauricio showed us footage of huge factory ships fishing right at Roca Partida.

'These big commercial fishing boats are setting nets trying to get yellowfin tuna,' he explained. 'They send helicopters up to spot them, and as soon as they see the tuna they set the purse seine nets.* These illegal boats catch tonnes of tuna right inside the marine protected area.'

'So what's the problem?' I asked rhetorically. 'Isn't there enough to go around?'

'No!' he said. 'Commercial fishing is devastating in MPAs because they get not just one fish but tonnes, plus bycatch like the silky sharks and the dolphins. This is the only place on earth that I've seen giant yellowfin tuna – and we've seen them breed here. These guys think they can do what they want and get away with it, but ecotourists are the eyes, we can take footage, take coordinates and we can call the Mexican navy. This is a national park, a no-take zone.'

However, the Mexican navy is too busy dealing with the drug, gun and people trafficking of the murderous Sinaloa Cartel, and simply cannot commit resources to illegal fishing in these seas.

---

* A method using a hanging net, primarily targeting species near the surface.

The one thing we had on our side on this occasion was Allejandro. As the director of the national park, he is the man with the power to make big change. In our few days here, Mauricio needed to present evidence to him that Clarion was vital, and that it was under threat. This crucial part of our plan was rather scuppered, though, when Allejandro got word on the satellite phone that he was needed in America on urgent legal business relating to the park. Without warning, he was whisked away on another boat, and we were all left wondering what on earth to do next.

The team held a council of war as we prepared to make land, clustered around the charts we had of Clarion island. The mapping that had seemed adequate when we were planning back home in the UK was now clearly woefully inadequate. Trying to identify aquatic features from these charts would be like trying to pinpoint the Croydon Odeon on a globe of the world. The known sites were at the extremes of the land-mass, where the currents and undersea geography would offer similar visuals to Roca Partida. It was extremely exciting but also rather intimidating, considering quite how much had not been surveyed and how challenging it was going to be to figure out the best places to dive.

In our council of war, we assigned tasks to the team members. The boat captain Julio would use his sonar to look for anomalies in the seabed. Obvious pinnacles and potential seamounts should show up in the substrate topography. Maru, our drone pilot, would fly high and put strong polarising filters on the lens of her camera to look down into the vodka-clear waters and point out submerged features that came close to the surface. Frida and our dive team identified and steered us clear of all known dive sites, and Mauricio rationalised spots he thought would be best for shark aggregations. The rest of us trained binoculars on the horizon for diving frigates, leaping tuna and dolphins, possible signs of feeding activity beneath the waves.

Powering down on the DPVs on our first dive was scary as hell. Granted, the tropical waters were clear, calm and warm, but there were still so many things that could go wrong. My underwater navigation is iffy at the best of times, let alone when powering along on a torpedo in a full face mask. Our plan was to run transects off into the blue, hoping to find something of significance. What we found was dolphins. Loads of them. Clarion had more bottlenose dolphins than anywhere I'd ever been, and they were incredibly inquisitive. They'd never seen DPVs before and couldn't resist them. One dolphin surfed along at the front of mine for about five minutes, his tail flapping in my face, threatening to knock off my mask.

It was immediately obvious that we could cover and map a really good distance on the DPVs. Our first dive allowed us to cover several hundred metres, our second half a kilometre. By our fourth dive a few days later, we travelled a whole kilometre underwater. Everything we saw, we logged and charted.

The most sublime geography could be found to the north-west of Clarion. A small island that looked like something out of Monument Valley was surrounded by smaller dark pinnacles and shards, all of which were being smashed by rolling white spume. The predicted high winds were inbound by this point, and we needed to get to the south side of Clarion, where we could shelter in the lee of the volcano. There was, however, time for one last dusk dive before we upped anchor.

We traversed the coast, staying below the crashing waves which churned the surface above us. The landscape was one of giant volcanic blocks, presumably tossed into the sea during an ancient eruption. Between them ran gullies filled with reef sharks and elongated trumpetfish, and on the outskirts countless green and olive ridley turtles rolled and paddled. Under the power of our scooters it felt like being a peregrine falcon, coursing through alpine valleys while hunting. Giant shoals of colourful tropical reef fish, parrotfish and butterflyfish flitted among the silver chub and bream of the temperate current = two distinct fauna (that of the tropics and temperate seas) co-existing in this unique place. It was the marine equivalent of finding grizzly bears and kangaroos drinking from the same stream.

The site was jaw-dropping, all the more so because we were the first recorded divers to see it. As such, it was our privilege to name it. We called it Roca Fregata, after the clusters of frigate birds perched on a black rock where we exited the water.

As the sun set and the swell gathered, the *Guardian* was under power again and bound for the southern coast in search of some meagre refuge from the weather howling in from the north. The expectation was that our next two days would be spent simply waiting here, drinking coffee and battling the incoming elements. We woke next morning though to find that lady luck was on our side. The 'low-lying' island of Clarion is anything but, and her hefty bulk provided an ample windbreak. The entire south coast was as flat as a mill pond, and we could see humpbacks blowing a mile away. Our sound recordist Nick put a highly sensitive hydrophone into the water, and we listened in awe as the whales sang to the sea, the most perfect and beautiful melody on earth. The only other boat was the illegal fishermen.

Just like at Roca Partida, the currents and the weather almost always seem to blow in from the north. That meant that the southern side was sheltered not just above but also below the waves. And it wasn't just humpbacks that nursed their calves in the warm, safe waters. It turned out to be a haven for others as well.

Once in the water, we headed west from our southeast starting point, keeping a set distance apart and powering on a compass bearing to cut as neat transects as possible. Again we had dolphins as our constant escorts, clearly enjoying gambolling in our wake.

With so much life evident, it was tragic to find balls of discarded thick fishing line tangled in much of the rocks. It seems likely that illegal boats like the *Royal Polaris* are often driven to shelter in the bay, and probably fish here pretty intensively. Mauricio suggested we move even further west, to an area where he had caught and tagged sharks in the past. He also had an acoustic receiver there he wanted to retrieve. As we neared the spot, Mark spoke on the underwater comms.

'Steve, hang on, check this out.'

I swam over to see what he had spotted. It was a cloud of yellow and black striped wrasse, and they were spawning. A female would rise up from the reef and drop her eggs in the water column, then five or six males would rocket up, eject their sperm and rocket back again. It was captivating. It wasn't, however, what Mark had noticed.

'It's a cleaning station!' he said. And he was right. Not only were there wrasse, but hogfish, Clarion angelfish and barberfish were all hanging out at one prominent rock, and several triggerfish, like colourful flattened rugby balls, were hanging, tails down and heads high, to be plucked clean of fish lice. Our new site was named Easter Island after the rock that loomed over it, which resembled one of the famous edifices.

Mauricio was particularly excited by the new cleaning station. 'We need to put down BRUVs here,' he said.

A BRUV, or baited remote underwater video, is essentially a camera stuck next to a bait box with some fish in it. They're not a cutting-edge tool, but much like camera traps on land, they allow you to watch and record while you go off and do other things. The added advantage is that there are no humans present to scare away the shy fish.

That night, Mauricio set us the task of trying to catch some of the sharks that use the bay. The recovery and processing of his data from the acoustic transmitter had obviously got him excited, and he was clearly working up a hypothesis in his head.

Sadly this was the first time in my life when I was not happy to see dolphins. For the entire night, a substantial pod of dolphins hunted squid and flying fish in the lights behind the boat, clicking and whistling and picking off every single fish that strayed into the light. For their part, the flying fish put on an elegant display. If they sensed the dolphins coming, they'd thrash their tails and leap before flying along the surface as if in slow motion. Some of them would properly take flight and lift off to more than head height. Some thwacked into the boat and knocked themselves senseless. One big one landed at my feet. In the hand, it looked like a half-starved mackerel, until you extended its wings – the pectoral fins. With a translucent crimson membrane between the struts, they looked like fairy wings.

With so many dolphins, Mau was never going to be able to find a shark, and we gave up at midnight. But there was still the question of what had been recorded on the BRUV cameras.

Frida, Mau and I sat around the computer and went back through the images. Many of them were simply inquisitive bottlenose dolphins sticking their smiling beaks near the camera. At one point an octopus billowed its skirts up into shot and sat there obscuring our view for about 15 minutes. But then Mau and Frida started to pick out sharks.

'Galapagos sharks!' Mau said. 'And they're babies, look!' He counted about ten of them. All tiny, no more than weeks old. This combined with the data from his acoustic tags left him in no doubt.

'This area south of Clarion is a nursery,' he said. 'The Galapagos sharks are coming from miles away – some have travelled 2,200 nautical miles to get here, from Galapagos, via

Clipperton Atoll. They come all this way to give birth, and then the little ones, they hang out here in the south of Clarion. They stay here a long time, maybe two years, before they even move to the north of Clarion.'

'And this is a highly migratory species, right?' I asked. 'I mean their nurseries are vital?'

'That's right,' he said. 'And we did not know where they gave birth before. It could have been like what we think with the hammerheads, that they head to the mainland and the mangrove to have their pups, or even like the whale shark going deep … but now we know. They come here.'

With that he looked across to the illegal fishing boat just a stone's throw away.

'Thing is, nature is very smart. These are top predators. They develop slowly, they reproduce slowly, they have low fecundity = some species have to wait 20 years before they can reproduce. They are not adapted to being fished like this. These people fishing here illegally, they will kill some of the pups, but they are also taking their food. If this is a sanctuary, and can stay a sanctuary, then there is hope for these sharks. And we saw baby silky sharks to the north too. Maybe they have nurseries here. The silvertips give birth and stay at San Benedicto. Blacktips migrate past here on their way up to the Sea of Cortez. This place is the key to their survival.'

It's very rare that you hear such candour and clarity from scientists. Often they are loathe to make big statements before they have published their papers. Mauricio, though, is a much more pragmatic conservationist. His demand was for nothing less than absolute protection for all waters where these sharks migrate, breed and pup.

This might seem extreme to some. After all, a shark is not a panda or a snow leopard, right? But tough times require tough measures. Despite the grand aims of 30 per cent by 2030 from the UN's new exciting High Seas Treaty, currently less than 3 per cent of our oceans enjoy any level of protection. The more our industrial fishing fleets strip mine our seas in a short-sighted search for short-term economic gain, the emptier our oceans will be, and the more tempting the no-take zones will become. Soon it will not just be bluefin tuna that command prices in the millions, but yellowfin, and then even the skipjacks and dogtooth, and then what? Total annihilation of our marine ecosystems for fun and profit. It can't happen. We need to expand our marine protected areas for the benefit of everyone, including fishermen, and we need to up our protection in line with the well-armed militia protecting every rhino in Southern Africa. The stakes are just as high, the outcome just as inevitable if we do not.

• • •

And what of the future? Nearly half the world's people live within the coastal zone, and 3 billion rely on seafood for their primary source of protein to survive. There will be more plastic in the sea than fish by 2050, and by then the UN estimates that the world's fish stocks will be so far gone that they will be commercially extinct. That's not the Pacific bluefin tuna, the orange roughy or the Atlantic cod … it's ALL our sea fish.

By then it's likely that all our tropical coral reefs will be bleached and dead. We already have 500 dead zones where nothing lives – an area bigger than the United Kingdom. We will need net carbon neutrality by then or face a rise in global temperatures that will sink a billion people, and lead to catastrophic conflicts and freak destructive weather events being the norm.

The year 2050 sounds like science fiction, right? But it's less than 30 years away. Go back in time 30 years, and I was about to start university. Time is relative. It rampages on despite our efforts to slow it down. My lifetime is everything to me, but it is nothing in the history of our planet. Extinction is the natural order of things – 99 per cent of species that have ever lived are extinct, the equivalent of at least 5 billion different kinds of animals. However, just as with global warming, cataclysmic changes such as we are seeing now are supposed to happen over tens of millennia, not over decades. And they are supposed to be driven by asteroids and giant volcanoes, not by the short-sighted short-termism of one species.

But if we are to address the gigantic problems of species extinction, climate crisis and human overuse of resources then the oceans also provide hope. Water is our panacea, the secret to wellbeing and wellness. The ocean is what makes our planet live-able – it gives us food, climate change mitigation,* and provides an energy and chemical buffer for the atmosphere. Since the Industrial Revolution the oceans have absorbed around a third of the carbon gases we've created and 90 per cent of the excess heat we've generated. Our ocean currents are responsible for our weather, our climate and the distribution of nutrients around the world. The Gulf Stream alone is 100 times bigger in volume than all the world's rivers combined. Sea plants like *Posidonia* create 70 per cent of the oxygen we breathe.

And there is so much left to discover. Hydrothermal vents, discovered not long after I was born, have revealed an entire ecosystem that was totally unknown to us and helped to unlock the greatest mysteries of evolution: how did we get here? How did life form on our planet? There are countless new discoveries to be made in the deep. Countless new species. Vast tracts of deep, dark void that hold secrets we can only dream of.

Blue carbon can save us; mangroves and seagrass and salt marshes can save us. Marine protected areas work staggeringly

---

* The oceans are able to hold 1,000 times more heat than the atmosphere.

well, and are easy to create (though much harder to police). Sanctuaries and no-take zones offer species a chance to rebound, even prosper, and pass on their bounty to surrounding seas. New tech and understanding can combine with the sheer magnitude of our oceans to provide the answers to all of our problems.

I have seen with these ageing eyes that if we give nature a chance to bounce back, she will. And I have also had the great privilege of working alongside people who have made quite phenomenal changes to save our seas. When my crew and I were long gone and onto our next projects, Mauricio was still working up his findings into scientific papers that were peer-reviewed and published. He presented his findings to Allejandro, the director of Revillagigedo National Park, who reacted with characteristic decisiveness. That same day, Allejandro sent a naval vessel out to patrol the less-trammelled regions of Revillagigedo, and decreed that he would find a budget for a full-time boat to patrol the waters around Clarion. Offenders fishing within the park limits would have their boats and gear impounded. Repeat offenders could face prison sentences.

And the data gathered by Randall Arauz and his compatriots in Galapagos, Ecuador, Panama and Colombia led to the ratification of the Eastern Tropical Pacific Marine Corridor, which encompasses all of their sovereign waters. The goal is to protect 800,000 square miles of sea, and to place all fisheries under improved management. It's an enormous effort, connecting dozens of NGOs and nations, but offers huge positive potential for regeneration. Galapagos too has expanded its marine protected area by 23,000 square miles, with its new sanctuary covering the migration routes of the same sharks we were studying.

In the UK in early 2023, we finally announced three underwater marine protected areas that will be no-take zones. And pledges made way back in 2009 at COP15 to protect 30 per cent of the world's seas seem finally to have brought forth fruit. The High Seas Treaty is the greatest conservation line in the sand of my lifetime and gives me enormous hope. We can stem the tide, and we must.

But we have to start with a truth. Our oceans are dying. No, enough with the passive language – we are killing our oceans, and this is the generation that has to act, so our children and theirs will be able to experience the wonders of the deep blue.

ACKNOWLEDGEMENTS

Huge thanks to Prof Steve Simpson for his stellar work on reef and fish acoustics, for introducing me to a whole other dimension of underwater life, for letting a lousy scientist like me be part of his research team, and most of all for being the kind of inspirational teacher every single science student should have!

Prof Boneboy 'Ben' Garrod. I'd love to take this opportunity to mock you for nerdery. But you've always been the instant answer to all my wildlife questions, assistance when my science is slack, passion when my enthusiasm wanes. You're the wind beneath my wings. Now go and play with your fossil collection.

To Alex, Randall Aruaz and the Fins Attached team on the Sharkwater. What adventures we had! And director of Mexico parks Alejandro Gonzalez for proving that protection can work, and can happen quick if the right people press the right buttons!

To Marteyne, Greg, Kiah, Jess, Miri, Pip, Julian and everyone else at the Blue Marine Foundation team from Six Senses Laamu, for giving me the opportunity to be a part of groundbreaking science in utter paradise.

Professor Peter Harrison and his team from James Cook University. Our time working together was probably the best holiday I've ever had!

Dr Mauricio Hoyos, we've had so many adventures together, and I've learned so much under your tutelage. Thank you so much for always 'getting' what we're trying to achieve, for bringing such professionalism to the game, and for letting us telly types steal all your research!

Acknowledgements

Scotty Carnahan. Get outside. Do stuff. It certainly wouldn't suck.

To José Antonio and medic Mike Hudson. For keeping me safe, and helping me to some of the finest moments in my life. Blue skies and fair seas. You will always be missed, my brothers.

For Steve Truluck, the crew of the Silurian and all the other homegrown orca nuts of Orcawatch. It's passionate pros/obsessives like you that make my job possible.

For Boone Hodgin at Ravenscroft, I so hope to be back with your glorious salmon sharks someday soon!

To big Si – writing this made me realise quite how much nuts stuff we've done together! To Marky Sharman (20 years and counting!), Bimini Dunc, Jonny Rogers, Rob T, Katy, Nicky boy and Parker, for being my crew and making it all look so good! And to Rosie, Brix, Sanna, Ruth, Tom, Ali, Ro, Emma (kitty), Wendy, Hannah banana, Petra and all our other crews who made most of these adventures possible.

Doc Sam Gruber and his teams (I could almost say disciples!) at the Bimini Shark Lab. I learned so much working with you. Thank you for being there at the very start of my education into the wonder of sharks, and for so much that came after.

Tanya Streeter for being a total inspiration, for your coaching, and making me aspire to a life aquatic!

To Jo Sarsby and the team at JSM, and Aimee too for living with my impossible calendar, making endless excuses for me missing Zoom calls and for having had my back and stuck with me for … well, forever. And Misty for your drive and enthusiasm. Love you, Miz.

Albert Petrillo, Julian Alexander and the team helped me turn this book from a pandemic dream into a reality. Thank you so much for sticking with me and it, and I so hope we can do more.

To Ali, Paul and all those at Shark Trust, Guy and everyone at Manta Trust, Gray and Suze at Bite Back and Alex at Fins Attached, who give selflessly to provide a better future for our marine life. Thank you for letting me be a part of your missions, and I promise to be better at getting to AGMs in the future!

To mine and Helen's family, for all your help and understanding. This would all have been impossible without you.

To my gorgeous kids, Logan, Kit and Bo. You are the answers to all the big questions in my life, the thing I was always searching for. I can't wait to find out who you're going to be, and for us to do some of these adventures together as a family. And to my wife, Helen, for still carrying on smashing it in the red, white and blue while being the best ever mum.

ABOUT THE AUTHOR

Steve Backshall MBE, PhD, Msc, BA hons, FRGS is the six-time BAFTA award-winning naturalist, who in 2020 was named 'Explorer of the Year' by the Scientific Exploration Society. His *Deadly 60* programmes are shown in 160 countries around the world. He is patron or president of 15 different wildlife, conservation and young person charities. Originally an author, then Adventurer in Residence for the National Geographic, Steve has visited 116 countries, described new species of animal, had a frog species named after him, and led first ascents of mountains, descents of unnamed rivers, mapped uncharted cave systems, discovered a 100-metre waterfall in the jungles of Suriname, and dived to take the first light into sunken cave passages in Yucatan. Steve is a professional diver and freediver, with more than 3,000 dives on every continent. He lives by the Thames in Berkshire, England with his wife, double Olympic gold medallist Helen Glover MBE, and their three children Kit, Bo and Logan.

INDEX

Index

Index

Index

Index

Index

Index